Astronomers' Observing Guides

Other titles in this series

Star Clusters and How to Observe Them
Mark Allison

Saturn and How to Observe It
Julius Benton

The Moon and How to Observe It
Peter Grego

Double & Multiple Stars and How to Observe Them
James Mullaney

Forthcoming titles in this series

Nebulae and How to Observe Them
Steven Coe

Asteroids and How to Observe Them
Lawrence Garrett

Venus and Mercury and How to Observe Them
Peter Grego

Jupiter and How to Observe It
John McAnally

Supernovae and How to Observe Them
Martin Mobberley

Total Eclipses and How to Observe Them
Martin Mobberley

The Messier Objects and How to Observe Them
Paul L. Money

The Herschel Objects and How to Observe Them
James Mullaney

Wolfgang Steinicke
Richard Jakiel

Galaxies and How to Observe Them

Wolfgang Steinicke
steinicke-zehnle@t-online.de

Richard Jakiel
rjakiel@earthlink.net

British Library Cataloguing in Publication Data
A catalogue record for this book is available from the British Library

Library of Congress Control Number: 2006926447

ISBN-10: 1-85233-752-4 Printed on acid-free paper
ISBN-13: 978-1-85233-752-0

9 8 7 6 5 4 3 2 1

Springer Science+Business Media

To my wife Gisela

–Wolfgang Steinicke

Preface

Galaxies have fascinated me since I started visual observations with a small 4 in. Newtonian reflector around 1966. Pretty soon all Messier objects were "checked off," and new targets had to be chosen. I marched through what might be called the "natural sequence" in the career of a visual observer: Messier, NGC, IC and UGC objects came out of the dark – glimpsed with growing apertures: 4 in., 8 in., 14 in., and finally 20 in. Over the years I've learned to be modest, concerning both targets and instruments. Each step in the sequence must be accompanied by a certain growth of knowledge concerning the physical nature of the targets.

I've also learned that blind faith in catalogues and their data can cause frustration. In the early days, it was not easy to get the relevant information. I was, for instance, fascinated by the entries in my old *New General Catalogue*: what's behind all these anonymous numbers? In my wildest dreams I wished to have access to the *Palomar Observatory Sky Survey*. In naked reality, however I must live with an old-fashioned sky atlas, showing stars to 7 mag, with a few galaxies plotted. Thus, to light up the dark, one has to be inventive! Over the years, using all kinds of articles and images available, numerous handwritten lists were created. Based on this stuff and ongoing observations, a more detailed picture of the sky and its objects could be painted.

This is long ago. Nowadays everything is childishly simple – and perhaps much less exciting! If you want to know for instance all about VV 150, switch on your computer, try Google, Guide, NED or ADS (you will later see what's behind these abbreviations), and pretty soon you will be covered with tons of data. Unfortunately, this does not automatically imply that you will be successful at the telescope. Technique, dark sky and a lot more is needed – not to forget experience!

It was in early 2003, when I got in contact with Mike Inglis, a professional astronomer, and author of some popular astronomy books, who asked me to write a book on "galaxies." It was easy to comprehend that this inquiry met my very interests! Thus it was only a matter of a few formalities before I started writing. And here is the result, which hopefully shows a bit of my affection for these, often inconspicuous, but always fascinating building blocks of the universe.

I would like to thank some people for their valuable support. First of all, I have to mention my wife Gisela, who contributed through her patience and valuable advice. Next are Mike Inglis, John Watson and Harry Blom who made it possible to write this book. Special thanks goes to Rich Jakiel – one of the most experienced observers in the United States – for his keen proof reading. He critically checked my text, concerning language, form and content. He also added some new aspects and information and nevertheless contributed many valuable observations.

Finally, I would like to thank other keen observers from all over the world, who offered their results for presentation. A large number of visual descriptions given here are based on their work. Particularly I would like to mention Steve Gottlieb and Steve Coe (both United States), Jens Bohle (Germany) and Magda Streicher (South Africa). The book presents a number of high-quality amateur astrophotos. These are due to Peter Bresseler, Werner E. Celnik, Bernd Flach-Wilken, Torsten Güths, Bernd Koch, Gary Poyner, Cord Scholz, Rainer Sparenberg and Volker Wendel. Hope to meet you all at the next star party!

Wolfgang Steinicke
November 2005

I grew up during the 60's and I fondly recall the excitement and high tension of the space race. It no doubt helped fuel my passion for the stars and I spent a great deal of time in the public library perusing the latest astronomy magazines and books. By the early 70's, I had become an avid star gazer, using a rusty old pair of 7 x 35mm Zeiss binoculars to explore the heavens from my backyard. In 1974, I got my first real telescope – a $4\frac{1}{4}$" Newtonian on a German Equatorial mount as a Christmas present. The first objects I saw were Jupiter, M42 and M31. I was totally hooked, and within a year I had seen several hundred new astronomical objects.

I quickly graduated to an 8-inch Cave reflector, which was to become my main instrument for the next ten years. With that relatively modest instrument, I observed nearly 2000 objects, and made detailed sketches of many of the brighter galaxies. Eventually, I moved up to using ever larger telescopes and my interest in astronomy deepened far beyond the mere observation of astronomical objects. Over the decades, I would observe thousands of galaxies, clusters, nebulae and double stars, plus write over 50 articles for a wide range of astronomical publications. This transition was in no doubt helped by the coming of the internet and vast online databases. I now had easy access to journals and references that were normally found in large university libraries. In time, I not only became interested in the structure of galaxies, but also their classification, formation and distribution in space.

In this lifelong astronomical journey, I've had a lot of help along the way. My mother was very instrumental in getting my "feet wet" in the sciences, through her gentle encouragement and many trips to the public library. Later on, Ernst Both (director of the Buffalo Museum of Science) gave me my first views through the telescope, and would become a life-long friend and mentor. I've also gained valuable experience, friendship and contacts as first a member of the Buffalo Astronomical Association (1980's), and later the Atlanta Astronomy Club (ACC). I'm still a very active member of the AAC, and fondly remember my many observing sessions with the "deepsky zombies". And finally, I'd like to give a big thanks to Wolfgang Steinicke for giving me the opportunity to first edit, and then add a number of new sections to this book. Co-authoring this book has been a very interesting experience and one I hope to repeat again in the near future.

Richard Jakiel

Contents

Section III **What to Observe? – The Objects**

Introduction

Undoubtedly, galaxies are among the most popular targets for the visual observer and they are a remarkably diverse class of deep-sky objects. In professional circles, galaxies are an extremely popular topic of research as the amount of scientific papers dealing with their structure, evolution, and cosmic significance is overwhelming. However, beginners are often disappointed when observing galaxies for the first time, due to their relatively inconspicuous appearance in the eyepiece. But realizing that the faint light has travelled millions of years in an expanding universe, or that an extragalactic monster emitted this feeble light at an early stage of the cosmic evolution, their reaction might be simply "Wow!" Thus, the observation of galaxies creates a feeling to be "involved" in one of the greatest mysteries of the universe. Beware that although a great deal is already known, many questions remain still open – and new mysteries arise, such as "dark energy."

We have attempted to address to all kinds of observers, with experience ranging from the novice to the seasoned veteran. This book presents an up-to-date collection of information and data. But it is neither a catalogue nor a mere list of observational data. It presents the necessary "theory" for visual observing galaxies by using a comprehensive collection of individual objects as representative examples. Though featuring the "visual" aspect, a critical comparison with photographic results might be always useful, but being aware that a beginner's perception is often heavily biased by "pretty pictures."

This book is divided into three sections. The first describes the physical nature, evolution and cosmic distribution of galaxies in their various forms and associations, as in pairs, groups, clusters or superclusters. All relevant astrophysical concepts and quantities will be discussed. An important theme, which is presented in the third part of this section, is the numerous – and sometimes confusing – catalogues and data, which open the door to individual objects. The observer will be introduced to the content, structure and reliability of classic and modern data sources.

Section II contains three parts. The first presents relevant information about useful accessories like finderscopes, eyepieces, or filters. Telescopes for visual observation, like the most prominent Dobsonian, are omitted, as they are described extensively in the literature. The second part is most important for visual observation, describing physiological and technical aspects: all about "exit pupil," "averted vision," or the relevance of "contrast and magnification" can be found. The third part deals with finding procedures, at which "starhopping" is favoured, and how to record, analyse or finally publish the observational results.

The third and most extensive section lists and describes a large number of sample objects. The simple question behind is: What to observe? The aim is to present various themes: from single observations up to complex programs. This arrangement reflects different aspects of galaxies. The objects are sorted according to certain categories: catalogues, sky areas, distance, appearance, higher-order structures, and finally some "odd stuff." The presentation is mostly "double": first the objects are listed with their individual data, followed by a section containing textual descriptions based on visual observations with different apertures. This might give a good idea of what to observe and what can be seen. Note that northern sky objects are dominant; nevertheless a number of southern galaxies, groups and clusters have been included. Though this section contains the bulk of the objects, many additional ones are mentioned in section I. As concerning their data, the standard catalogues or sky mapping software should be consulted.

The appendix presents a collection of general literature, like books, magazines, or printed sky atlases, and digital sources, like sky mapping software, Internet databases and other important websites. You may wish to consult these first two parts of the appendix when such sources are mentioned in the text. All other references, like books and articles, mainly of a special kind and only relevant at the specific place in the text, are designated by a number in brackets. This refers to the large collection listed in the third part of the appendix. Note that actual articles or those from popular magazines (e.g. *Sky & Telescope*) are favoured. Primary sources, which appeared in the professional journals (e.g. *Astrophysical Journal*), are mentioned only if necessary. An index was omitted. The detailed Table of Contents, further sub-titles in the text, and the information given in the appendix make it easy to direct the reader to the subjects. To list all objects, mentioned in the tables was too expensive.

Finally here are a few technical notes on notation found throughout the book. Equatorial coordinates refer to the standard equinox J2000.0; units (like "h, m, s") are generally omitted. In the tables, constellations are referred by their common abbreviation, e.g. UMa for Ursa Major. Aperture is given (traditionally) in inch (in. or ″), or in metric units (cm, m); 1 in. = 1″ = 2.54 cm. Wavelength is measured in nanometers; 1 nm = 10^{-9} m. Distance is measured in light years (ly), or megaparsecs (Mpc); 1 Mpc = 3.26 Mill. ly. Other abbreviations are listed in the appendix.

Section I

Galaxies, Cluster of Galaxies, and their Data

Galaxies and clusters of galaxies are certainly among the most popular targets for amateur astronomers. They show an incredibly diverse range of size, shape, and internal structure has undoubtedly lead to their fascination among both amateurs and professional astronomers alike. However, this sheer complexity of form and evolution makes it necessary to discuss in detail the physical nature of galaxies and their place in the cosmic hierarchy. This first section outlines some of the current information on these objects. It concentrates on the current astrophysical facts relevant for observation, including catalogs and data. A few sample objects are presented to illustrate some of the major points. The rest of the objects will be presented in more detail in the last section of the book.

Chapter 1

Galaxies, Cluster of Galaxies, & their Data

Galaxies are vast aggregates of stars, dust, and gas ranging from a few thousand to nearly a million light-years in diameter (see e.g., the classic book of Hodge). Their respective masses show a similarly broad range from less than a million to well over trillion solar masses [1]. This variety of shape and form is far greater than in any other class of deep sky objects – often demonstrated in close vicinity (Fig. 1.1). But visually galaxies often appear as only a small, diffuse patch of "light" in a small telescope – rather mundane and subdued especially when compared with the brighter open clusters, galactic, and/or planetary nebulae. However, when viewed with a moderate to large sized scopes, many of the brighter galaxies will reveal a wealth of detail to the seasoned observer. The delicate swirls of the spiral arms may be detected, along with smaller structures as bright knots and dark rifts and lanes. But even then don't expect to see in the eyepiece anything similar what is present on photographic images! Photography (especially if in color) and visual observation are different worlds. Starting the observing career with galaxies thus might cause some initial frustration. Visually galaxies are shy targets, which must be handled with care. Using the right equipment and learning good observing techniques are valuable in their study.

Nevertheless, galaxies are most popular targets for many reasons. First, there is the enormous distance involved: galaxies are truly cosmic objects populating deep space (see Waller & Hodge). To be visible over millions of light-years, they must produce an incredible output of energy. By far the most extreme are the quasars – so luminous that they are visible (even in amateur telescopes) at distances of 10 billion ly [2]. Many galaxies are now known to host a central supermassive black hole, which appears to be key in powering the cores of the most active examples [3,233]. Another important characteristic is their tendency to form pairs, groups, and clusters. Often many different types of galaxies are associated with these clusters making them rich targets for study (Fig. 1.2). In the dense environment of large clusters the gravitational interaction between the member galaxies is a common process and can produce a variety of unusual tidal phenomena.

Galaxies are the building blocks of the universe. Their creation and evolution has essentially defined the large-scale cosmic structure [4]. Over decades, astronomers have measured the recessional velocity of galaxies known as the *redshift* (interpreted as cosmic expansion) to produce a three-dimensional picture of the large-scale structure in the universe. Since light travels with a finite speed, everything we observe has happened in the past. With the largest telescopes, astronomers are able to observe the conditions of the remote past. Galaxies are late witnesses of the big bang [5,6], which happens 13.7 billion years ago. Shortly after this initial "burst" of creation of the universe, the first structures appear, triggered by large amounts of cold dark matter. Formed by gravity and angular

Fig. 1.1. Galaxy pair NGC 5090 and NGC 5091 in Centaurus

momentum, clouds of primordial hydrogen and helium slowly fragment into smaller portions ("protogalaxies"). Early star formation and gravitational coalescence eventually convert them into the "first" true galaxies. We now know that the development of galaxies strongly depends on gravitational interactions in the small early universe.

To sum up: it is the extreme physical nature, the significance as building blocks of the cosmos, and the variety of forms and interactions, that makes the study of galaxies a fascinating topic. It is those few photons entering our eye, after traveling millions of years through space-time, are enough to create the special "galaxy feeling."

The Milky Way and the Nature of Galaxies

Our Host Galaxy: The Milky Way

We live in a galaxy, called the Milky Way [7,227]. Unfortunately, being observers inside the system, we are not able to observe our galaxy as a whole. This is much like trying to "see the forest through the trees." The primary reason for most of these problems is interstellar absorption. Obscuration from interfering clouds of dust and gas make it very difficult to penetrate in visible light. A short view from outside would be enough to realize the major facts about the structure and dynamics of our galaxy. Fortunately, the interstellar matter is pretty transparent for radio and infrared radiation. It took some time of

Fig. 1.2. The rich galaxy cluster A 1656 in Coma Berenices

applying sophisticated astrometric, statistical, and spectral methods to our galaxy and studying external galaxies to reach our current state of knowledge (Fig. 1.3). A classic source is the book by Bok & Bok.

Not only the internal view is reduced, but the dense dust bands of the Milky Way also block parts of the cosmic scenery. This area of the universe, dimmed in the optical spectral range has been nicknamed the "zone of avoidance" (ZOA) by 19th century astronomers. But this dusty veil is quite uneven and some galaxies do shine through some of the thinner regions [8]. Fortunately, the unobscured part of the sky is much larger, presenting a tremendous number of extragalactic systems for observation and study. Over the past 100 years, huge strides have been made in galactic astronomy and we now know a great deal about the structure and evolution of the Milky Way and other galaxies [9]. For example, the nearest large galaxies are the Andromeda Nebula M 31 (Fig. 1.4) and the Triangulum Nebula M 33. We have learned that both are not mere neighbors but very similar systems: spiral galaxies, of comparable in size and composition that are dynamically related to our own system.

The Milky Way is estimated to be at least 10 billion years old. Our galaxy is in many respects a quite ordinary galaxy and is thus used as a standard – similar to the sun, which defines a standard for stars. With a mass of at least 180 billion solar masses it is a fairly large, but otherwise unremarkable spiral galaxy. But the Milky Way is by no means an

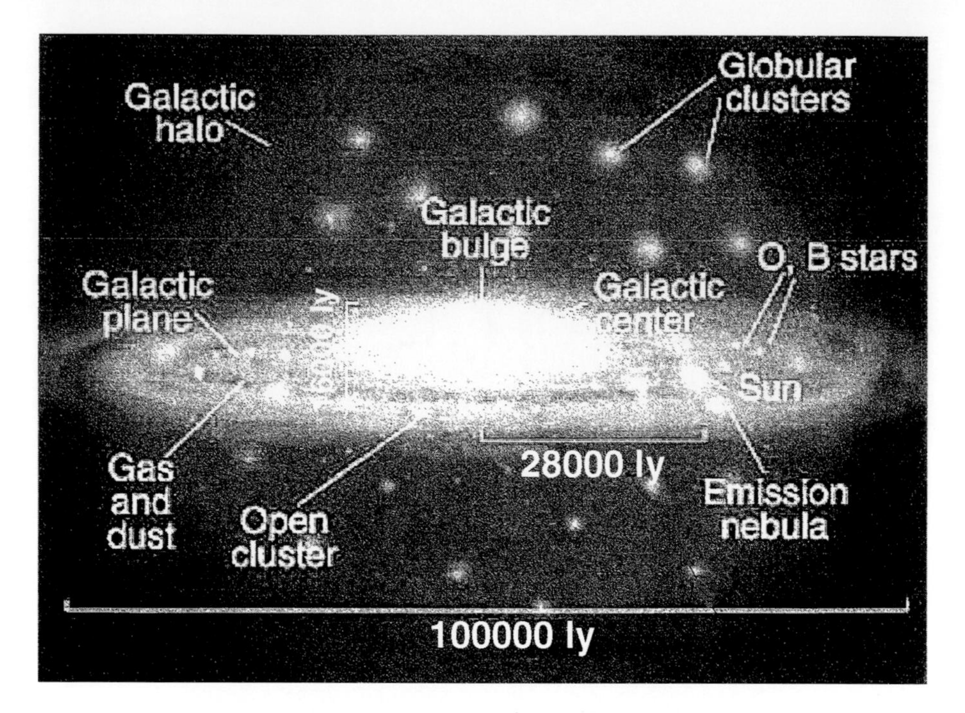

Fig. 1.3. Structure of the Milky Way

aging diva, it is a dynamic object, showing a continuous regeneration [253]. About 10% of the *visible* mass is in the form of dust and gas, while the rest is distributed in stars and nonluminous bodies. Since most of the stars are less massive than the sun (only a small fraction is heavier), a "true" star count would result in a much higher number.

The most prominent feature is the disk, about 100,000 ly in diameter but only 16,000 ly thick. It is not uniform, but divided into several spiral arms, which contains the bright, young stars and most of the interstellar matter. This structure is what the ancients called the "Milky Way" – a broad, diffuse glowing band that encircles the entire sky. The irregular distribution of dust, gas, and stars produces large local variations in brightness. As often the case – many of the bright areas such as the Scutum cloud are also intermixed with dark, heavily obscuring *molecular clouds.* Perhaps the most prominent of these is the southern "coal sack" (Fig. 1.5).

The disk encloses a central region, the nuclear bulge. While our neighbor, the Andromeda Nebula M 31, is an ordinary spiral galaxy with a spherical center, the Milky Way seems to be a barred spiral, i.e., the bulge is (slightly) bar shaped. This was recently confirmed by a University of Wisconsin team using NASA's Spitzer Space Telescope [243]. Visually we can get only a rough impression of this region in the form of a concentration of bright star clouds in the direction of Sagittarius. Details are obscured by large amounts of dust. What we know about the central part of our galaxy comes from radio and infrared radiation, which is much less absorbed. The galactic center is a strong radio source, called Sgr A. It hosts an extremely compact, supermassive object, the black hole Sgr A*. We are not unique in hosting such a gravitating monster, many galaxies including nearby M 31 have them residing in their core regions. The nucleus of our galaxy is extremely small and is not optically visible even though it is packed with hundreds of giant stars orbiting the black hole at high velocities. From our location of nearly 28,000 ly distance – it appears as a tiny condensation only 1″ across, which corresponds to a linear diameter of 10 ly.

Fig. 1.4. The nearest large galaxy: M 31 in Andromeda

Disk and bulge of the Milky Way are surrounded by a large spherical halo of faint stars, 600,000 ly in diameter. It is the home of the globular clusters, revolving the center in elliptical orbits of high eccentricity. At present over 150 of these objects are known. In competition with M 31 the Milky Way comes off as second best as our giant neighbor has at least twice as many. Some of our globulars might be hidden by the ZOA, but the present theory favors a number of less than 200.

Broken down into the basic elements – our Milky Way consists of 73% hydrogen, 25% helium, and 2% "metals" (in astrophysics all elements heavier than helium are called "metals"). This matter is roughly distributed as: 10% bright stars (being also the most massive), 80% faint stars (the sun is among them), 10% gas, and 0.1% dust. These fractions differ significantly when looking at individual structures, e.g., bulge, disk, or halo. The bulge and the globular clusters contain mainly old stars, called "population II." These stars are metal-poor, having only one-tenth of the metallicity of our sun. Population II defines the first generation of galactic stars. They contain primordial matter (hydrogen, helium), still not polluted with heavy elements, created later in massive stars and supernovae, and injected in subsequent generations of stars. These old stars survived due to their low mass, which causes an economic consumption of their fuel. Even older are the one billion halo stars, which may have been created 600 million years earlier than the Milky Way itself.

Fig. 1.5. The southern Milky Way with the "coal sack" in Crux

The disk contains the young stars, called "population I" (there is a more detailed population scheme, not needed here). The spiral arms are still the cradle of new stars. Here the raw material needed for star formation is available in the numerous molecular clouds. Through the process of gravitation accretion the gas and dust is condensed into stars of different mass. Often a large number of stars are created at once, building an open cluster. Unused interstellar matter is often visible in the vicinity of young luminous stars. In case of HII regions, hot stars ionize the hydrogen atoms, which emit photons of red light when recombining. Such structures are generally called emission nebulae (Fig. 1.6).

If the star is not hot enough or too far away to ionize the gaseous part of the interstellar matter, one may see a reflection nebula. The dust reflects mainly the blue light (Fig. 1.7), though they may also be yellowish in color. Dust absorbs starlight and such areas may be visible as dark "rifts" or "holes" against the bright stellar background. All such types of galactic nebulae, present in the spiral arms, are closely related with star formation. The youthful population I stars are metal rich compared with the much older population II. They belong to subsequent generations, containing heavier elements which are created by nuclear fusion processes in red giants or during a supernova explosion. Massive stars live a very short life as they convert hydrogen into helium at a prodigious rate. At present the star formation rate in the disk is around 1 star per year. This does not explain the several hundred billion disk stars. Undoubtedly the rate of stellar formation was significantly higher in the past.

The sun is located in the outer half of the disk, about 28,000 ly from the center. Compared to the dense, chaotic central region, the outer disk is a much better place to observe the Milky Way and the rest of the universe. The Milky Way offers a variety of interesting objects, located in the nearby spiral arms [10], e.g., open clusters, planetary nebulae, or emission nebulae. Our local spiral arm is called "Orion arm" (Fig. 1.8),

Fig. 1.6. The bright HII region IC 5146 ("Cocoon Nebula") in Cygnus

Fig. 1.7. Reflection nebulae NGC 6726/27/29 in Corona Australis

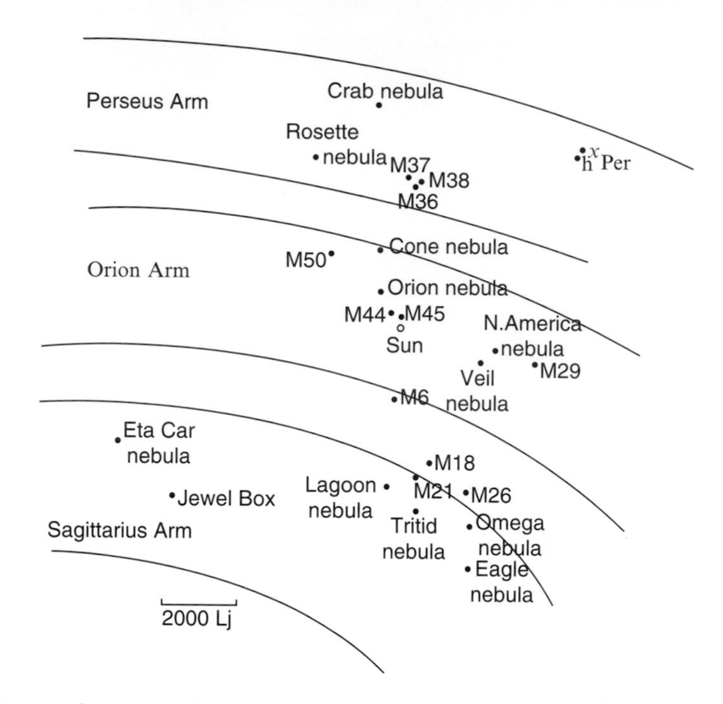

Fig. 1.8. Local and neighboring spiral arms with a sample of embedded nebulae and clusters

containing the young belt stars of Orion and the Orion Nebula, 1,600 ly away. It is also the home of the open clusters M 6, M 29, and M 50, and the planetary nebulae M 57, M 27, and M 97. The next outer arm, the "Perseus arm," contains the three Auriga clusters M 36, M 37, and M 38, and supernova remnant M 1, the Crab Nebula some 6,300 ly away. The next inner arm is called "Sagittarius arm," highlighted by the emission nebulae M 8, M 17, and M 20 (Trifid Nebula; 5,200 ly) and the bright open clusters M 18, M 21, and M 26. This region is also in the same direction of the galactic core.

The flat disks and the spiral structure of galaxies like the Milky Way strongly suggest some kind of rotation. Basically all gravitational systems, lacking inner forces (like radiation pressure in a star), must show some kind of movement to be stable. A good example is our own solar system. The galactic rotation can be detected from the earth as relative motions of the stars. Unfortunately, stars show also individual (peculiar) motions. Both effects combine on the sphere to the "proper motion." The human eye cannot detect this, as star positions remain unchanged in a lifetime – thus the term "fixed star." By accurate measurements (comparing precise star positions from different epochs) proper motion becomes evident. However, even the nearest stars show shifts of only a few arc seconds per year. To study the real space motion, the radial velocity is needed, derived from the Doppler shift of the spectral lines of the star. These space velocities can be some 100 km/s. The problem is to filter out the part due to galactic rotation. By "stellar statistics," where thousands of stars are measured, and radio astronomical methods (spectral shifting of the 21 cm-line of neutral hydrogen) the rotation curve of the Milky Way can be determined. The main result is that the Milky Way's rotation velocity depends on the distance from the galactic center. It first increases, slows down a bit to become nearly constant in the outer disk (Fig. 1.9). At the position of our

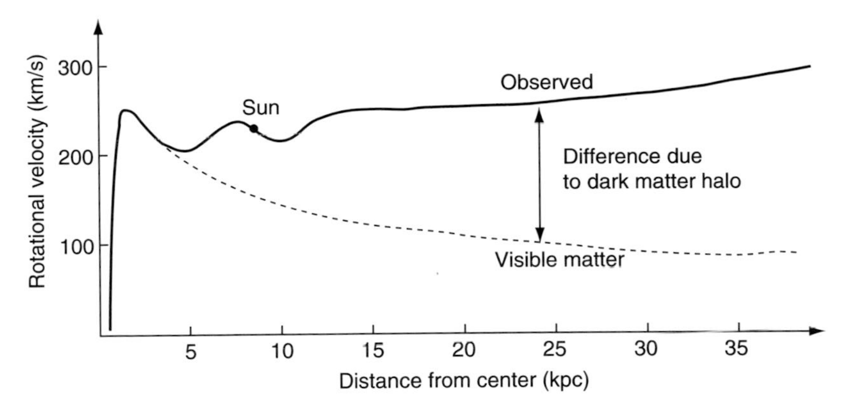

Fig. 1.9. The observed rotation curve of the Milky Way shows strong evidence of dark matter

sun, the velocity is 220 km/second, leading to a rotation period of 200 million years, called a "galactic year."

The form of the rotation curve bears a fundamental problem – not only for the Milky Way, but also for spiral galaxies in general. Taking into account the visible (luminous) matter, e.g., stars or hot gas, the velocity must decrease significantly in the outer disk. But the measured values show no such decrease. This requires far more matter than what is currently observed. Without it, the system would be unstable, throwing out stars by the centrifugal force. The amount of the "missing mass" is immense: the total mass of the Milky Way must be six times higher than the observed mass in form of luminous matter. What is the nature of the "dark matter" and where is it located? A possible place is the galactic halo; populated by faint, low mass stars and perhaps invisible brown dwarfs. We will see later that this is not a satisfying solution for the mass deficit of spiral galaxies.

Parameters of Galaxies

To describe the main features of galaxies, a few parameters are necessary. Similar to stars, their values show a great variety. The appearance of galaxies depends both on physical and geometrical characteristics. We therefore distinguish between these interior and exterior parameters.

Interior parameters reflect the astrophysical properties of the galaxy: linear dimension, mass, luminosity (absolute magnitude), rotation, and content (stars, interstellar matter). They mainly describe the overall features, thus may also called "integral quantities." There is a more or less strong relation between them, e.g., rotation and mass. The measurement of such quantities is a difficult problem, being not directly observable. For example, to determine linear dimension or absolute magnitude, the distance (not an interior parameter) must be known. The morphology of the galaxy can give valuable hints on its astrophysical properties, thus various classification schemes were developed.

Exterior parameters reflect geometrical properties: position (coordinates), distance, spatial orientation (position angle, inclination, elongation). We may add apparent brightness and angular diameter here, which depend on distance and interior parameters (luminosity, linear diameter). With the exception of distance, all these parameters are directly measurable.

Redshift and Distance

Assuming – in a first approximation – "given" similar sizes and luminosities for galaxies, then nearby galaxies will appear large and bright, distant ones small and faint. Comparing similar types of galaxies, this rule is helpful for an initial estimate. In case of individual objects the error can be pretty large: as known from stars there are also dwarfs and giants among the galaxies. The determination of reliable extragalactic distances is therefore a complicated task. A series of overlapping methods with different precisions, the "cosmic distance ladder" (Fig. 1.10), must be applied [11,12,206,215]. Crucial steps on the ladder (distance indicators) are Cepheids and RR Lyrae stars, bright stars (e.g., luminous blue variables), globular clusters, bright HII regions, novae, and supernovae of Type Ia. Other methods use the Fisher–Tully or Faber–Jackson relations, and the Zeldovich–Sunyaev effect (see below). Besides using light-years, extragalactic distances are often measured in Megaparsec (Mpc), where 1 Mpc = 3.26 million ly.

Cepheid variables are luminous pulsating stars. Due to the celebrated period–luminosity relation it is possible to calculate the absolute magnitude by measuring the period of the light variation. A comparison with the apparent magnitude then gives the distance of the star. Cepheids are frequent in galaxies and can be detected with the aid of the *Hubble Space Telescope* (HST) up to distances of 100 Mpc.

To determine the distances of a large number of objects or if there is no other reliable distance indicator, there is a practicable method: the "redshift" of the galaxy. The crucial tool is the "Hubble law," in that:

$$v = H_0 r.$$

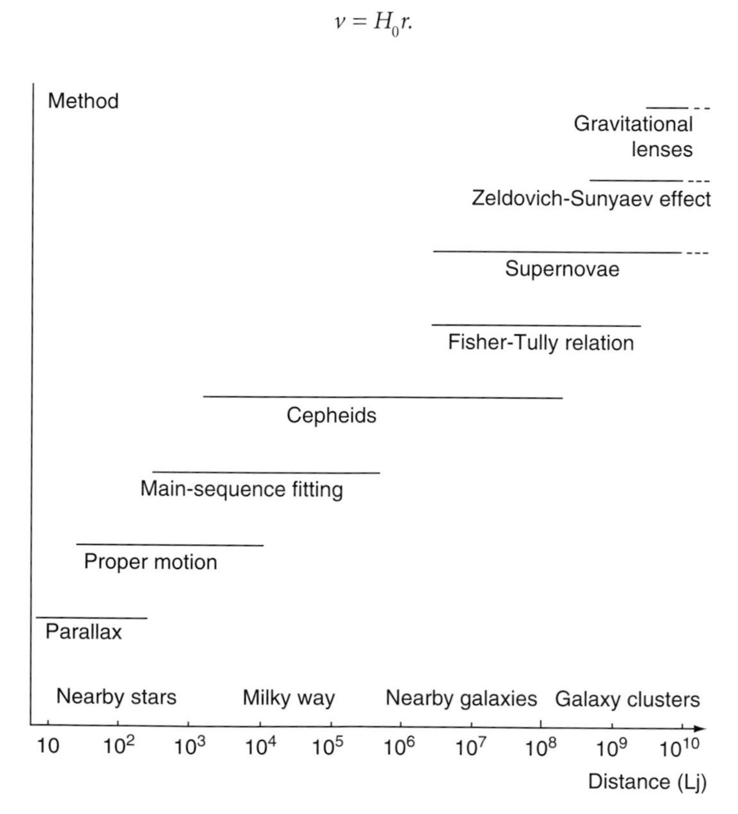

Fig. 1.10. The cosmic distance ladder based on overlapping methods

It states that the measured "radial velocity" v is proportional to the distance r. H is the proportionality factor called "Hubble parameter" (it is not a constant, since it changes with cosmic time). The main problem is to calibrate this relation, e.g., to determine the present (local) value of the Hubble parameter (indicated by the index "0"). This was made (which much controversy) using the cosmic distance ladder, but the latest value is based on satellite measurements of the cosmic background radiation, giving $H_0 = 71$ (km/s)/Mpc.

To get the distance r, the radial velocity v has to be measured. It results from the shift of spectral lines in the spectrum of the galaxy. What causes this shift? The Doppler effect states that the spectral lines of an emitter (e.g., hot gas) are collectively shifted to the red if the source is moving away from the observer or to the blue if approaching. Let λ be the measured wavelength of a spectral line (e.g., hydrogen) and $\Delta\lambda = \lambda - \lambda_0$ its shift (difference between measured and labor value), then z is defined by $z = \Delta\lambda / \lambda$. The Doppler effect gives the relation $z = v/c$ (c = velocity of light), thus the shift is proportional to the velocity. Most galaxies show a redshift due to a "recession velocity." A few nearby ones, like the Andromeda Nebula, show a blueshift, thus approaching us. The Hubble law has been confirmed to distances of billions of light-years (Fig. 1.11). Looking back in time, when the universe was smaller, H_0 roughly determines its age. Using this relationship astronomers can derive an age, which is a bit higher than that of the oldest stars or globular clusters.

Be careful with the idea of a "recession velocity" for galaxies as implying a certain kind of motion. In terms of Einstein's General Relativity, the Hubble law is a consequence of the expansion of the universe [13,254]. Galaxies take part in this expansion. But only space grows, not bound systems, like human bodies or galaxies – otherwise we would not detect any expansion since all objects including the measuring rods would grow in an

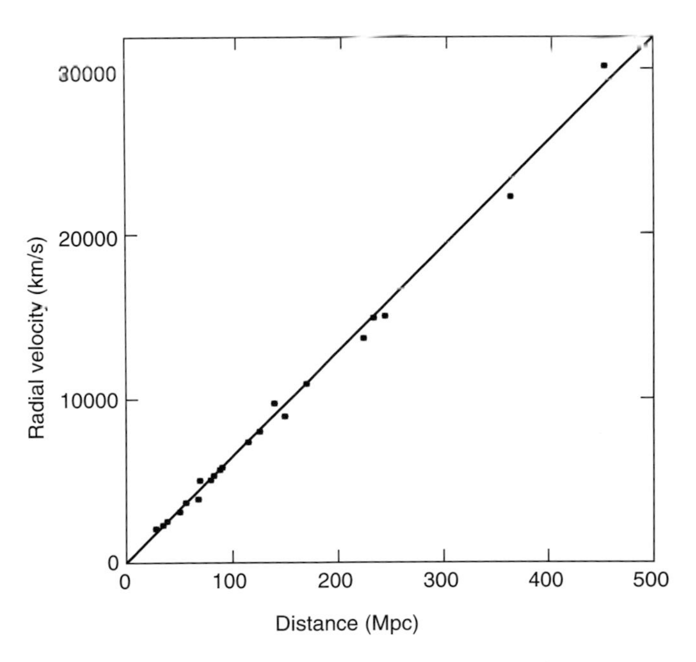

Fig. 1.11. Hubble's law is verified up to great distances

equal manner. Recession velocities are an illusion: There is no dynamical motion in cosmology. The galaxies are "fixed," merely carried along by the growing space – like points on the surface of a balloon, which is uniformly blown up. In terms of General Relativity, redshift is caused by a "cosmological" Doppler effect, which has nothing to do with radial velocity: the expansion stretches the light to a longer wavelength [14,15]. At present, general relativistic cosmology has turned a corner with exacting measurements of the expansion and evolution of the universe [16].

Nevertheless, in case of galaxies one uses the term "radial velocities" as a synonym for redshift. This is not totally wrong. Redshift, being the primary observable quantity, does not by itself give any hint where it comes from. Indeed it can contain a fraction due to real dynamical motions, locally induced by gravitational forces. One can imagine that such "peculiar motions" of galaxies are the main reason for the problems and controversies in determining the local Hubble parameter. In case of the Andromeda Nebula, the gravitational attraction by the Milky Way (and vice versa) dominates the expansion, the net effect is a blueshift. Another prominent peculiar motion is the "Virgo flow," caused by the gravitational pull of the Virgo Cluster on the galaxies of the Local Group. Fortunately, the significance of peculiar motions in the redshift decreases with larger z. At greater distances expansion the smooth "Hubble flow" always wins the race!

Position, Elongation, Position Angle, Inclination

Coordinates

In contrast to the third dimension (distance), the spherical coordinates (right ascension, declination) are much easier to determine. As the basic reference frame is oriented on the celestial equator, we talk about "equatorial coordinates" [17]. The right ascension is abbreviated R.A. ("ascensio recta"); the formula letter is α and the units are hour, minute, second. Right ascension runs from 0 to 24 hours (west to east). Note that east is to the left on the sky, while it is to the right on an atlas of the earth. The origin is defined by the vernal equinox. Declination is abbreviated "Decl"; the formula letter is δ and the units are $^\circ$ $'$ $''$ (degree, arcminute, arcsecond). Declination runs from -90° (south celestial pole) via 0° (celestial equator) to $+90^\circ$ (north celestial pole). The two axis of a parallactic mounted telescope (hour axis, polar axis) naturally follow these coordinates. Note that the scales of α and δ are not equal: at the celestial equator we have $1^m = 15'$. Thus right ascension should be written with an extra digit for equal accuracy. Writing 12 34.5 +06 27 is correct, but 12 34 +06 27 is not. Toward the celestial poles the scale difference decreases; for $\delta = 80^\circ$ there is $1^m = 2.5'$.

The direction of the Earth polar axis is not constant, but displays a slow, but complex motion known as precession and nutation. Thus the equatorial reference frame is time dependent. The "wobbling" Earth affects the orientation of the celestial equator in space and therefore the position of the vernal equinox. This leads to a passive change of the coordinate values (α, δ) of any celestial object. To become independent of the date of observation (epoch), one uses a coordinate system, which refers to a fixed date, the "standard equinox." It is defined by the beginning of a certain year, e.g., 1900, 1950, or 2000. At present the standard equinox is J2000.0, referring to the position of the celestial equator at the beginning of the (Julian) year 2000. It follows B1950.0 (B means "Besselian year").

Equatorial coordinates referring to different equinoxes (which must be always indicated) show different values. Coordinates can be "precessed" by formulae to any standard (a common feature of sky mapping software).

Position is not only a matter of coordinates. In case of stars we must consider the proper motion. The position must then refer to the date of measurement. Fortunately, galaxies show no such motion on the sphere. With the exception of quasars, galaxies are extended objects. The positional accuracy depends on defining a center. For "normal" types like spiral or elliptical galaxies, this is obvious, but in case of large, irregular, or asymmetric systems it is not. To define the very center is difficult or even arbitrary. In such cases, like IC 1613, NGC 4861, or NGC 55 (Fig. 1.12), the literature offers different coordinates, sometimes with a senseless degree of precision. It is always useful to denote the point, e.g., a bright condensation (knot), to which the measured coordinates refer.

The "horizontal system," defined by azimuth and altitude (elevation), depends on the location on earth. Altitude is the angle above the horizon, running from 0° (horizon) through 90° (zenit); azimuth is the horizontal direction, from south (0°) via west (90°), north (180°) to east (270°). Horizontal coordinates are essential for the local visibility of celestial objects. Another system is "galactic coordinates," used for objects in the Milky Way. "Supergalactic coordinates" fit to galaxies in the Local Supercluster.

Angular Size, Orientation

Position is a crucial parameter for identifying a galaxy, others are brightness (apparent magnitude), angular size, and position angle. In case of angular size one usually gives the larger and smaller diameter (a, b in arcmin) of an ellipse roughly covering the object. Most galaxies are elongated ($a > b$), mostly due to orientation, but sometimes it

Fig. 1.12. The asymmetric galaxy NGC 55, a member of the Sculptor group

represents the actual shape. The orientation of elongated objects on the sphere is measured by the position angle (PA), which is the angle between the north direction and the larger axis (a). This value ranges from 0° (north) via 90° (east) to 180° (south). If a and b are nearly equal, the position angle cannot be given with certainty; in such cases the literature cites different values.

In case of flattened systems (spiral galaxies) one defines the inclination i, which gives the orientation of the galaxy (disk) in space. It is the angle between the rotation axis (perpendicular to the disk) and the observer, varying from 0° ("face-on") to 90° ("edge-on"). Note that inclination is opposite to the "tilt angle" between the plane and the observer (see de Vaucouleurs [257]). Depending on inclination, different structures become visible. Prominent face-on galaxies are M 83, M 101, and NGC 1232 (Fig. 1.13), present their spiral arms and star formation regions for easy viewing. If they show two large, well-defined arms the term "grand design spiral" is used. Even in the intermediate case of M 51 ($i = 45°$) the spiral pattern is easily visible. A more difficult case is M 31, with $i = 72.5°$, which is fairly edge-on. With higher inclination, spiral- and bar structures become hidden, but other features like a prominent bulge and/or the equatorial dust band are easy to see. The reason for the latter is the internal extinction by opaque interstellar matter (dust). Examples of edge-on galaxies are M 104 ($i = 84°$), NGC 4565 ($i = 86°$; Fig. 1.14), NGC 5907 ($i = 86.5°$) and NGC 891 ($i = 88°$).

Position angle, inclination, and (orientation-based) elongation are directly observable quantities of an accidental nature. If definable (good examples are spiral or lenticular galaxies), they are purely geometrical and not related with interior parameters. The inclination of elliptical galaxies is not obvious, being naturally elongated systems.

Fig. 1.13. The face-on spiral galaxy NGC 1232 in Eridanus

Fig. 1.14. The edge-on spiral galaxy NGC 4565 in Coma Berenices

Apparent Magnitude, Angular Diameter

The quantity "magnitude," as a measure of brightness (both terms are often used synonymous), comes in many different forms: One speaks about B-, V-, integrated, total, photographic, or surface magnitudes. To understand these values, e.g., given in galaxy catalogs, one should be familiar with the relevant definitions. Note that there is no uniformity, even concerning the units used. Thus different data sources are not easily comparable. Maybe the following explanations help to clear the situation [18,19].

Integrated and Total Magnitude, Surface Brightness

In principle, brightness comes in two opposite ways: from point and extended sources. Point sources, like stars or quasars, cause no trouble as the brightness is naturally concentrated (integrated) at a point. But an "integrated magnitude" can also be defined for extended objects like galaxies. In this case one thinks of the incoming light as being concentrated (focused) into a point, to be compared with the magnitude of a reference star. The integrated magnitude can be determined with a photometer, where the radiation is focused on the detector. Actually the intensity is measured, which differs from the brightness. The brightness is proportional to the logarithm of the intensity and that's how our eye responds to light: in a roughly logarithmic fashion.

Talking about magnitude, one usually means "integrated magnitude," abbreviated *m*. Its unit is "mag," writing $m = 13.5$ mag for instance; also in use is ^{m} (don't confuse this with "minute"). Comparing different magnitudes, a smaller value refers to a brighter source; mathematically it is 9 mag < 10 mag, but 9 mag is brighter than 10 mag!

The concept of "surface brightness" (SB) is quite opposite. It is defined for extended objects by the apparent magnitude per (spherical) surface unit, usually abbreviated with m'

and measured in mag/arcmin^2 or mag/arcsec^2. The value difference is 8.89, i.e., 10 mag/arcmin^2 is equal to 18.89 mag/arcsec^2. Every surface unit has a specific brightness, think e.g., of a "pixel" in a CCD image. Giving $m' = 13$ mag/arcmin^2 for a galaxy says, that a $1' \times 1' = 1$ arcmin^2 fraction shows a brightness equal to a 13 mag star. To get a visual impression, how bright (or better: faint) this looks, use a high magnification eyepiece and defocus a 13 mag star to a patch of $1'$.

Surface brightness is calculated in "dividing" the integrated magnitude by the area covered by the object (see formulas below). One usually gets an average surface brightness, which is a suitable measure only for objects showing a more or less homogeneous brightness distribution, e.g., compact galaxies. Bright galaxies (like M 33 or M 82) show details of different surface brightness. Thus the average does not represent the real situation. We will later see that surface brightness is an essential quantity for visual observing, while the mere integrated magnitude often tells not much about visibility.

A magnitude usually refers to a standard "color." The UBV-system defines magnitudes in the near ultraviolet, blue, and visual (yellow) part of the spectrum. For measurement the photometer is equipped with a standard filter with peak transmission at 365 nm (*U*), 440 nm (*B*), or 550 nm (*V*). In addition, there are *R*-(red) or *I*-(infrared) magnitudes, defined at 700 and 900 nm, respectively. To assign the part of the spectrum used, one writes e.g., m_B or m_V (alternatively *B*, *V*) in case of the integrated magnitude and m_B' or m_V' (alternatively B', V') for the surface brightness. Normally the *U*-, *B*-, and *V*-magnitudes of a galaxy are different. This leads to the definition of color indices: $B–V$ or $U–B$. For most galaxies it is $B > V$, they are fainter (!) in the blue, than in the visual (yellow) light. Typical values of $B–V$ are: 1.1 for elliptical galaxies, 0.7 for spiral galaxies, 0.4 for irregular galaxies, and 0.0 for "blue compact galaxies" (BCD). Quasars show a variation between 0.0 and 1.0 while Seyfert galaxies are around 0.5.

Total and Standard Magnitude, Standard Diameter

To measure the brightness of galaxies one often uses a diaphragm. The value of the integrated magnitude depends on its aperture. With increasing size, the magnitude rises to reach saturation, representing an "infinite" aperture. This limit is called "total magnitude," abbreviated B_T or V_T (Fig. 1.15).

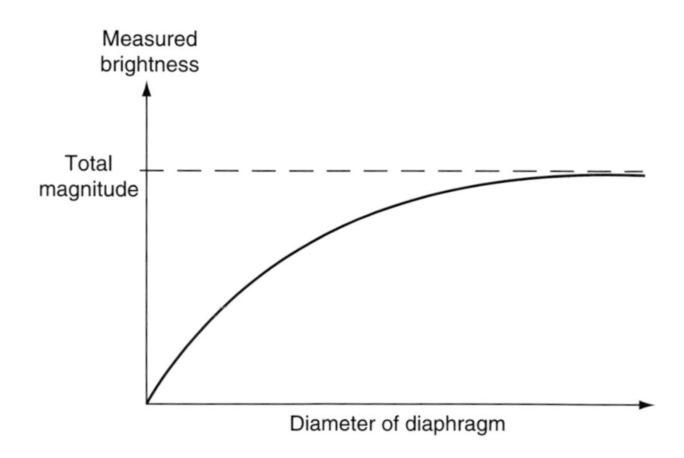

Fig. 1.15. Definition of total magnitude (see text)

The "standard magnitude" is an integrated magnitude too (usually in B), but it needs the surface brightness for definition. By plotting isophotes, i.e., lines of constant surface brightness, the "edge" of a galaxy can be defined by the "standard isophote" at a level of 25 mag/arcsec^2 (in B). This corresponds to 1/10 of the night sky surface brightness. The ellipse-shaped standard isophote defines the "standard diameters" a_{25} and b_{25}. Also used, but a bit larger, are the *Holmberg* diameters, as defined by the isophote at the 26.5 mag/arcsec^2 level. The integrated magnitude inside the standard isophote is called "standard magnitude" B_{25}. It is equivalent to around 90% of the total B-magnitude (B_T). One can calculate the (average) surface brightness inside the standard isophote by the following formula:

$$B'_{25} = B_{25} + 2.5 \log (a_{25} \cdot b_{25}) - 0.26.$$

It uses the standard *B*-magnitude and the standard diameters in arcminute. The term "0.26" converts the rectangular area into an ellipse. Often the (standard) input parameters are not present. If only the total visual magnitude V_T and a not specified size (a, b) is available, the following formula roughly gives the average visual surface brightness:

$$V' = V_T + \Delta + 2.5 \log (a \cdot b) - 0.26.$$

The term Δ *corrects* the standard into the total magnitude. It is 0.25 for elliptical galaxies, 0.13 for lenticular galaxies, and 0.11 for spiral galaxies. If V_T is not available, but B_T and $(B-V)_T$, one can calculate $V_T = B_T - (B - V)_T$. If a galaxy catalog lists *surface brightness*, usually a calculated value of V' (or B'_{25}) is implied. Note that for small galaxies ($a \cdot b < 1$), the surface brightness gets significantly higher than the integrated magnitude. This is most extreme for almost stellar objects, like quasars. In this case the quantity "surface brightness" makes no sense at all.

Photographic Magnitude

In contrast to the preceding definitions, photographic magnitude (m_{pg}), as used in some catalogs, is a weakly defined quantity. It usually corresponds approximately to a *B* magnitude. Often it only declares that the magnitude was determined from the density on a film. *O*- and *E*-magnitudes refer to the two versions of the first *Palomar Observatory Sky Survey* (POSS), using blue-sensitive Kodak 103aO-plates, and red-sensitive 103aE-plates.

Despite the fact that galaxies are extended objects, how do photographic magnitudes arise in galaxy catalogs? Generally, galaxies on the plate must be compared by a certain method to the density caused by reference stars of known magnitude. Shapley and Ames used for their *Survey of External Galaxies Brighter than the 13th Magnitude* (1932) [20] a wide-angle lens. Even large and bright galaxies created an almost point-like image. In comparison with modern B_T magnitudes, the Shapley–Ames magnitudes show an error of 0.5 mag, thus are not very reliable. Nevertheless they have been used in many popular catalogs and handbooks, e.g., *Burnham's Celestial Handbook.*

The reverse way was taken by Zwicky in his *Schraffiermethode*, created for the magnitude system of his *Catalogue of Galaxies and of Clusters of Galaxies* (CGCG). The idea was to bring about a definite tracking error while exposing a plate with the 18″-Schmidt camera on Mt. Palomar. This smears out star images to an area of 1′. By visual inspection of the images, galaxies can be compared with reference stars of known magnitude – a time

Fig. 1.16. NGC 6946 a face-on spiral, obscured by the Milky Way in Cygnus

consuming, but pretty effective method! For calibration Zwicky uses the Shapley–Ames magnitudes. One easily guesses that his magnitude system gets not much better. Nevertheless many modern catalogs, e.g., the *Uppsala General Catalogue* (UGC), use Zwicky's photographic magnitudes. Relative to the modern B_T system, they are too faint by 0.3 mag, the difference to V_T is around 1 mag. This can be positive for visual observing. At low galactic latitudes, e.g., in regions obscured by the Milky Way, which cause a reddening of the light, a galaxy can be visually more than 2 mag brighter in comparison with the Zwicky magnitude. Some galaxies, listed as faint, turn out to be promising targets. Examples are NGC 6946 (m_{pg} = 10.5 mag, V = 9.0 mag; Fig. 1.16) and IC 10 (m_{pg} = 13.5 mag, V = 11.8 mag).

Classification

In the past, the incredibly diverse appearance of galaxies was a major roadblock to their understanding. A way to bring order into the chaos of their optical appearance is through various classification schemes [21]. The observed shapes and structures of galaxies reflect internal astrophysical properties and are a key to understand their evolution.

Hubble Classification

In 1926, Edwin Hubble revolutionized our basic understanding of galaxies by introducing a new classification system. According to Hubble, galaxies can be divided into the following types: elliptical (E), lenticular (S0), spiral or barred spiral (S, SB), and irregular (I). His original two-dimensional classification (Fig. 1.17) was known as the "tuning-fork" scheme. The types on the left (starting with E0) were called "early" by Hubble, the right ones (ending with Sc, SBc) "late." As we now know, this does not correspond with an evolutionary sequence [22].

The Hubble classification has been subject to many revisions and extensions documented by Sandage in two monumental works. The first was the *Hubble Atlas of Galaxies* which features (with minor modifications), the classic Hubble scheme. Whereas the 2-volume *Carnegie Atlas of Galaxies* introduces a more extended system, and is illustrated by a larger set of examples. Another revision was produced by de Vaucouleurs, based on a system already developed in the *Second Reference Catalogue of Bright Galaxies* (RC2). Many galaxy catalogs, trying to include the best available data, present a mix of different classification schemes. For faint galaxies often only a rough differentiation into "S" or "E" is given.

Elliptical galaxies normally show a symmetric shape. The surface brightness decreases smoothly from to center to the outer parts. The ellipticity ranges from E0 (round) to E7 (highly elongated). The true figure can be prolate, like a cigar or spindle (e.g., NGC 741), oblate, like a thick biconvex lens (e.g., NGC 315), or even triaxial and lacking any symmetry [23]. A typical E0 galaxy is NGC 5898 in Libra. An example of E6 is the "Spindle Galaxy" NGC 3115 in Sextans, while NGC 4623 in Virgo is of the rare type E7. Most ellipticals are pretty featureless objects, but a few show a box-shaped body or weak absorption structures. Examples of the latter, which often are prolate systems, are NGC 1947, NGC 5266 (Fig. 1.18), and IC 4370.

The type S0 ("lenticular," sometimes denoted "L") defines the transition between elliptical and spiral galaxies. S0-galaxies are lens-shaped systems, normally without any spiral structure

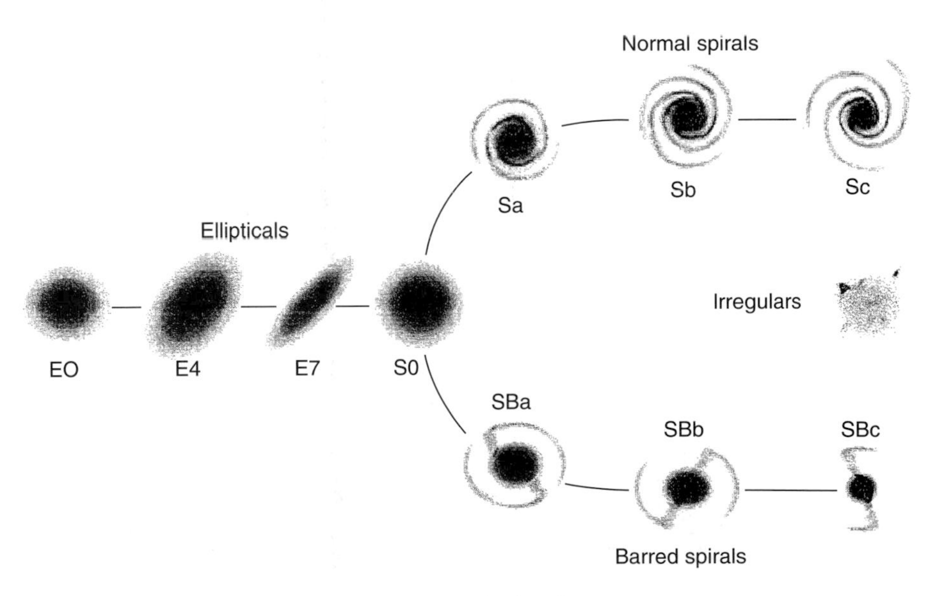

Fig. 1.17. Hubble classification

Fig. 1.18. The prolate "dusty" elliptical NGC 5266 in Centaurus

and a typical example is NGC 4111 in Canes Venatici. They are less flattened than the disks of spiral galaxies, showing a pretty high surface brightness. In some cases dust bands or even a weak bar structure (type SB0) are present. The central region of S0-galaxies is spherical, similar, but less massive than the bulges of spiral galaxies. There are also peculiar objects with a box-shaped center, e.g., NGC 128 in Pisces (Fig. 1.19) or NGC 7332 in Pegasus. This feature is also known from the study of elliptical galaxies. To resume, lenticular galaxies have connections to both elliptical and spiral galaxies thus are well placed by Hubble.

Common features of ordinary spiral galaxies (S) and barred spiral galaxies (SB) are the bulge (the spherical central region), and the surrounding disk containing the spiral arms. Ordinary spirals show two or more spiral arms starting smoothly from the outer bulge. The bulge of SB galaxies is bar shaped, with two spiral arms branching off. The prototype of a barred spiral is NGC 1365 (Fig. 1.20).

To classify spiral galaxies, the following features are commonly used:

- the form and density of spiral arms
- the relative dimensions of bulge and disk ("bulge-to-disk ratio")
- the form of bulge: spherical for type S, bar-shaped for type SB

The first two criteria are equivalent. This fact is useful, as sometimes only one is applicable. In case of high inclination (edge-on), the density of spiral arms is not visible; instead the bulge-to-disk ratio can be easily determined. How to see a bar in this case? This is rather difficult, but sometimes the spiral arms are disconnected above or below the disk at the end of the bar. An example is the SBc galaxy NGC 7640 in Andromeda.

Fig. 1.19. NGC 128 in Pisces, a lenticular galaxy with box-shaped center

Fig. 1.20. The prototype of a barred spiral: NGC 1365 in Fornax

The classic Hubble scheme was extended beyond the (late) types Sc and SBc. Galaxies of type Sd or SBd consist of a disk with extremely wide spiral arms and a very small bulge. No bulge is present for types Sm or SBm ("Magellanic" systems); the spiral structure is nearly lost. The following Table 1.1 describes the classification criteria in detail. For further differentiation, intermediate types, like Sab (NGC 4826) or SBdm (NGC 4236), can be applied.

The transition from Sm and SBm to irregular systems (I, Irr) is small. According to Holmberg's system, there are two types of irregulars: Irr I ("magellanic") and Irr II ("peculiar" or "amorphous"). Interestingly the Magellanic Clouds, originally classified as Irr I, are now representing type SBm. In case of the Large Magellanic Cloud (LMC) the bar structure with spiral arms is clearly visible on small scale images. Examples for type Irr I are dwarf galaxies, like IC 1613, a member of the Local Group, or the bright starburst galaxy NGC 4449 in Canes Venatici. The really chaotic are Irr II-systems. The prototype is M 82 (Fig. 1.21), other examples are NGC 520, NGC 3077.

Table 1.1. Classification of spiral galaxies

Spiral arms	Bulge-to-disk ratio	Bulge: spherical	Example	Bulge: bar	Example
Narrow	High (dominant bulge, small disk)	Sa	M 104	SBa	NGC 7743
Intermediate	One (bulge and disk comparable)	Sb	NGC 2841	SBb	M 95
Wide	Low (small bulge, large disk)	Sc	M 101	SBc	NGC 253
Very wide	Very low (tiny bulge, disk dominant)	Sd	NGC 7793	SBd	NGC 4242
Magellanic	Zero (no bulge, disk only)	Sm	NGC 4449	SBm	IC 2574

Fig. 1.21. The amorphous galaxy M 82 in Ursa Major

De Vaucouleurs Classification

In contrast to the simple Hubble scheme, the de Vaucouleurs classification is a multidimensional (Table 1.2). It first appeared in the *Second Reference Catalogue of Bright Galaxies* (RC2) and was finally established in its successor, the RC3 [24]. There are different levels of specification (class, family, variety, stage); whereas "class" is comparable to the Hubble type. In case of spiral galaxies the class is divided into the three "families": S, SAB, SB. While S and SB denote ordinary and barred spirals as in the Hubble system, SAB is a mixed case. Another dimension is "variety": (s), (r), and (rs), describing how the spiral arms fit to the bulge. In case of (s) they rise immediately from the bulge, leading to an S-shaped spiral pattern. If they start tangential to the bulge, forming an inner ring structure, we get case (r); (rs) describes a mixed situation. Finally the "stage" is the familiar subdivision into a, b, c, d, and m. The other classes (elliptical, lenticular, and irregular galaxies) show similar differentiations. Features titled "additional" must be set; to choice are *c* and *d* for elliptical galaxies as well as the stages (applicable for all types) in the last row. The symbols (R) and (R′) are prefixes, *pec* and *sp* are suffixes; the symbols ":" and "?" are inserted at the relevant place.

Two designations should be explained: cD and cI. Galaxies of type cD are gargantuan systems, having an elliptical-like nucleus surrounded by an extensive envelope. They are located in the centers of galaxy groups or clusters. They can weight well over a trillion solar masses, hosting gigantic central black holes [2]. Some objects show multiple nuclei; examples are NGC 6166 (Fig. 1.22) in A 2199, or ESO 146-5, the central galaxy in A 3827, where a hundred "Milky Ways" could easily fit inside. Such systems are cannibals, having devoured other galaxies in their lifetime.

The type cI describes the opposite extreme: compact intergalactic HII regions, also called "blue compact dwarfs" (BCD). They show a rapid star formation; I Zw 18 (PGC 27182) and II Zw 40 (PGC 18096; Fig. 1.23) are prominent examples [25].

The de Vaucouleurs classification is much more complex than Hubble's. It offers a great potential, but requires more work – and the galaxy must show enough features (a problem for faint, remote objects). The examples in Table 1.3 show, how to use the various features (see Fig. 1.24 for an example of a cE0 galaxy). The further development of classification schemes is subject of current research [255,256].

The general superiority of de Vaucouleurs' classification scheme against Hubble's can be demonstrated best with prominent galaxies (Table 1.4).

Classification depends on the experience of the classifying person. In case of spiral galaxies another factor is the detected radiation (spectral band). In the infrared or UV spiral galaxies look pretty different compared to visible light, some features can be weaker, others stronger – or even new features appear [239]. This can have influence on the appearing type [26,27]. For a complete classification based on galaxies observed with the *2 Micron All Sky Survey* (2MASS) see [28].

Special Cases, Peculiarities

The de Vaucouleurs classification offers symbols to denote abnormal structures, but must fail for really peculiar objects ("pec" does not tell much in this case). We now know that most peculiarities are due to interaction. Let's take a look at some oddities in the extragalactic zoo.

Still pretty normal is a "warped disk," often found in spiral galaxies. In this case the disk appears not flat, but twisted like the brim of a hat. Most likely this is due to a close encounter with a much smaller galaxy, ending up in a merger. A beautiful example is the

Table 1.2. Galaxy classification according to de Vaucouleurs (adopted from RC3)

Class	–	Variety	Stage (additional)
Elliptical ("E")		Compact cE	
		Dwarf dE	
			Ellipticity...E0
			thru...E6 (allowed:...E1.5)
		Special variety: cD ("core dominant")	–
Class	**Family**	**Variety (additional)**	**Stage (additional)**
Lenticular ("S0")	Ordinary SA0		
	Barred SB0		
	Mixed SAB0		
		Inner Ring S..(r)0	
		S-shaped S..(s)0	
		Mixed S..(rs)0	
			Early S..(..)0^{-}
			Intermediate S..(..)0°
			Late S..(..)0^{+}
Class	**Family**	**Variety (additional)**	**Stage (additional)**
Spirals ("S")	Ordinary SA		
	Barred SB		
	Mixed SAB		
		Inner Ring S..(r)	
		S-shaped S..(s)	
		Mixed S..(rs)	
			S..(..)0/a
			S..(..)a
			S..(..)ab
			S..(..)b
			S..(..)bc
			S..(..)c
			S..(..)cd
			S..(..)d
			S..(..)dm
			S..(..)m
Class	**Family**	**Variety (additional)**	**Stage (additional)**
Irregular ("I")	Ordinary IA		
	Barred IB		
	Mixed IAB		
		S-shaped I..(s)	Magellanic I..(..)m
		Special Variety:	Non-magellanic I..(..)0
		Compact cI.. ("blue compact dwarf")	–
			Stage (all types)
			Peculiarity pec
			Uncertain :
			Doubtful?
			Spindle sp
			Outer ring (R)
			Pseudo outer ring (R′)

Fig. 1.22. The multicore cD galaxy NGC 6166 in Hercules, dominating the rich cluster A 2199

Fig. 1.23. The "blue compact dwarf" II Zw 40 in Orion

Table 1.3. Examples of de Vaucouleurs classification

Classification	Description	Example
dE1	Dwarf elliptical with low ellipticity	IC 3501
cE0	Round, compact elliptical	NGC 4486B
E6	Elliptical with high ellipticity	NGC 5028
SA0°	Ordinary, intermediate lenticular	NGC 984
SB(rs)0+	S-shaped barred late lenticular with inner ring	NGC 5076
(R′)SA:(s:)a	S-shaped ordinary (both uncertain) spiral of type a with pseudo outer ring	IC 5075
(R′)SA(r)b	Ordinary spiral of type b with inner ring and pseudo outer ring	NGC 3329
(R)SAB(rs)0/a	S-shaped mixed spiral (transition from S0 to Sa) with inner ring and outer ring	IC 1895
SB(s)c	S-shaped barred spiral of type c	NGC 7573
SB(rs)b? sp	S-shaped barred spiral of (doubtful) type b with spindle form	NGC 4718
IABm	Mixed irregular of magellanic type	UGC 5510
I0 pec	Nonmagellanic irregular with peculiarity	IC 2458
(R′)IB(rs)m	S-shaped barred irregular of magellanic type with inner ring and pseudo outer ring	MCG -4-52-27
cI pec:	Compact irregular with uncertain peculiarity	MCG 3-5-13

Fig. 1.24. NGC 4486B, an example of type cE0. This extremely compact galaxy is a close companion of M 87 in the Virgo Cluster

Galaxy	Hubble	de Vaucouleurs
Milky Way	Sb	SAB(rs?)bc
LMC	Irr I	SB(s)m
SMC	Irr I	SB(s)m pec
M 31	Sb	SA(s)b
M 33	Sc	SA(s)c
M 82	Irr II	I0 sp
M 85	S0-a	SA(s)0° pec
M 104	Sa	SA(s)a sp
M 110	E5	E5 pec
NGC 253	SBc	SAB(s)c
NGC 1365	SBb	SB(s)b

Table 1.4. Classification of prominent galaxies in the Hubble and de Vaucouleurs schemes

Sa galaxy ESO 510-13 (Fig. 1.25). It seems interesting that both the Milky Way and the Andromeda Nebula are (weak) examples as well!

The term *superthin galaxy* [29] assigns an extremely flat type of galaxies, associated with the types Sc, Sd, or Sm. Superthin galaxies are unusually thin, featureless disks. Such objects are underdeveloped systems with an extremely low star formation rate. Through the lack of gas and dust, there is almost no internal extinction. Examples are NGC 100 (Fig. 1.26), IC 2233, UGC 7321, or UGC 9242. Due to the flat shape, an eventually warped disk can be easily detected, as in the case of UGC 7170, or even more extreme: UGC 3697, the "Integral Sign Galaxy" in Lynx.

Compact galaxies appear nearly stellar. The nucleus is dominant, surrounded by a weak, diffuse halo. Due to their blue color, many of these galaxies where first cataloged as "blue stellar objects" (BSO). Most of them are classified as *active galactic nuclei* (AGN) [30], which are related with quasars. This term designates galaxies with a high luminosity and emission line spectrum. The Seyfert galaxies M 77 (Fig. 1.27) and NGC 4151 are prominent examples. Many more can be found in the catalogs of Markarian (Mrk), Zwicky (Zw), or Haro. The activity comes from the core hosting a central black hole.

The term "peculiar" is used for galaxies, showing a wide range of abnormal features. It was first applied to galaxies by Halton Arp in his celebrated *Atlas of Peculiar Galaxies*,

Fig. 1.25. ESO 510-13 in Hydra, a spiral galaxy with a warped disk

Fig. 1.26. The superthin galaxy NGC 100 in Pisces

describing many objects already found by Vorontsov-Velyaminov (VV). Among the most spectacular cases are ring galaxies, originating from an impact of a small elliptical galaxy on a large spiral. The collision was central or orthogonal to the disk. After the penetration, a ring (actually the rest of the disk) and the two nuclei are left. Examples are the "Cartwheel," NGC 985 (VV 285), NGC 7598, or II Zw 28 (PGC 16572). Most ring galaxies are more or less deformed, but a few are highly symmetric: a small nucleus with a circular ring. One of the best examples is "Hoag's Object" (Fig. 1.28); similar cases are NGC 6028, CGCG 127-25, and MCG -2-33-25.

Ring galaxies should not be confused with the ring structures (s) and (R) mentioned in the de Vaucouleurs classification, which are not due to a collision. Similarly exotic and rare are "polar ring galaxies," consisting of a cigar-shaped galaxy with a central outer ring. A prominent example is NGC 4650A in Centaurus.

Facing such diversity, it is not surprising that many objects were wrongly classified. They can thus simultaneously appear in catalogs of entirely different object classes. Some errors were corrected, mainly due to a much improved measurement technique, but some still remain in the catalogs.

Integral Parameters and Evolution

The appearance (classification) of a galaxy is helpful to describe the internal properties. But to truly understand the astrophysics, quantities like luminosity, mass, linear dimension, and composition must be known.

Fig. 1.27. M 77 in Cetus, a prominent Seyfert galaxy

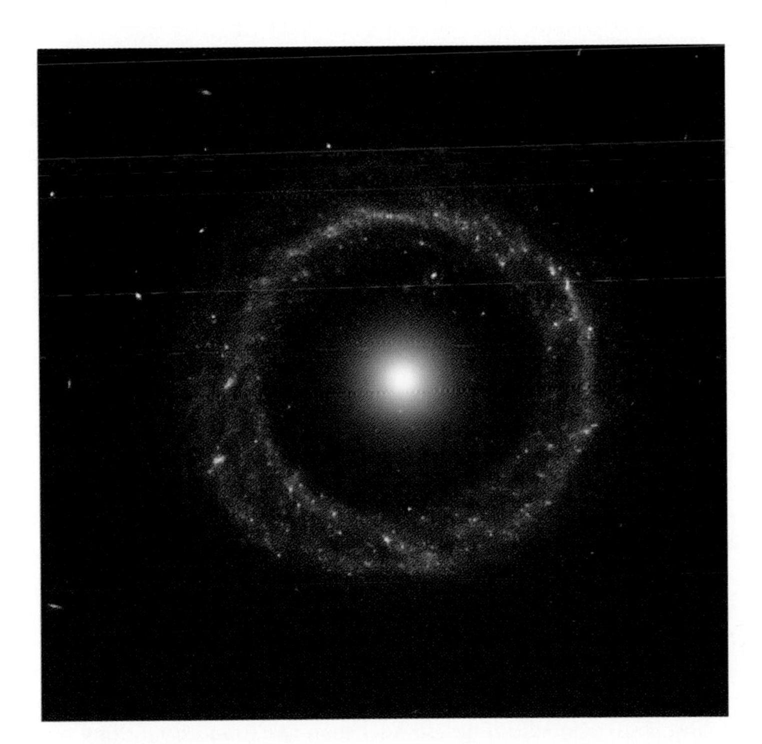

Fig. 1.28. Hoag's Object in Serpens, the perfect ring galaxy

Luminosity, Mass, and Rotation

Luminosity L (measured in Watts) is equivalent to absolute magnitude M, which is derived from apparent (integrated) magnitude and distance. The absolute magnitude of an object is the magnitude it would show in a standard distance of 10 pc = 32.6 ly. The absolute magnitude of the sun is M = 4.7 mag (sometimes ^{M} is used for the unit).

As for stars, the luminosity and mass of a galaxy are interrelated. The luminosity results from the sum of their luminous matter (stars, hot gas). We have already mentioned that in case of spirals, there is much less luminous matter than gravitating matter. The relation can be quantified by the mass-to-light ratio or M/L. What is the M/L for spiral galaxies? The gravitating matter is strongly related with rotation. Only if centrifugal and gravitational forces are equal (at any point), the system is stable. The rotation velocity of a spiral galaxy can be derived via the Doppler effect of the spectral lines. This can be done best by using galaxies with a large inclination (M 31 is already a good candidate). The part of the disk approaching us shows a blueshift that recessing a redshift (in the rest frame of the galaxy). Note that this kinematical Doppler shift has nothing to do with cosmological redshift.

Spiral galaxies, like the Milky Way, show a nearly constant rotation velocity of up to 300 km/s over a large fraction of the outer disk. To explain this unusual behavior, a large amount of unseen (dark) matter is needed. For the Milky Way we have $M/L \approx 6$, i.e., there is six times more "dark matter" than luminous (visible) matter. For spiral galaxies in general, M/L ranges from ~10 (for type Sa) to nearly 2 (for type Sm). For a single galaxy, this ratio increases with distance from center, thus the amount of dark matter needed gets higher toward the edge. This leads to the assumption that most of the "missing mass" is located in a dark halo around spiral galaxies, called the corona, which shows an M/L of 100 to 1,000.

What kind of matter is it? It can be shown that ordinary (baryonic) matter like dwarf stars, brown dwarfs, or isolated planets cannot explain the deficit. Another possibility are neutrinos, but are "hot dark matter" (fast moving) and thus not suitable. Favored are exotic, yet undetected kinds of matter, like axions or supersymmetric particles (e.g., the neutralino). It is now generally accepted that only 5% of the mass of the universe is due to "ordinary" matter. The epoch-making insights of Hubble and Einstein led us believe in an explainable "standard" universe [31]. But we are now strongly forced to quit our "familiar" universe: the major part or nearly 95% turns out to be a complete mystery!

The rotation of spiral galaxies yields a relevant contribution to determine distances. It causes a Doppler broadening of the 21 cm-line of neutral hydrogen. The line width (defined by the FWHM value = "full width at half maximum") is correlated with the absolute magnitude of the galaxy (Fisher–Tully relation; Fig. 1.29). A great line width, thus a large rotational velocity, corresponds to a high luminosity. When compared with the apparent magnitude of the object, it is possible to derive a distance.

While disk stars are moving in orderly orbits, bulge stars behave like a swarm of bees. This is the characteristic behavior of an elliptical galaxy as they have a "pure bulge" population. These are sometimes referred as "hot" stellar systems as most of the support against gravitational collapse comes from random ("thermal") motions, rather than ordered (rotational) motion. Individual motions cannot be determined, but instead we use global quantity called the "velocity dispersion." It is the range of velocities in the line of sight, smearing out spectral lines by Doppler shifting (like mentioned above). The velocity dispersion of elliptical galaxies can reach 400 km/s. A few systems show a weak general rotation, with velocities much smaller than the random motions. This rotation causes a slight flattening. Depending on the rotation velocity, the shape of the elliptical

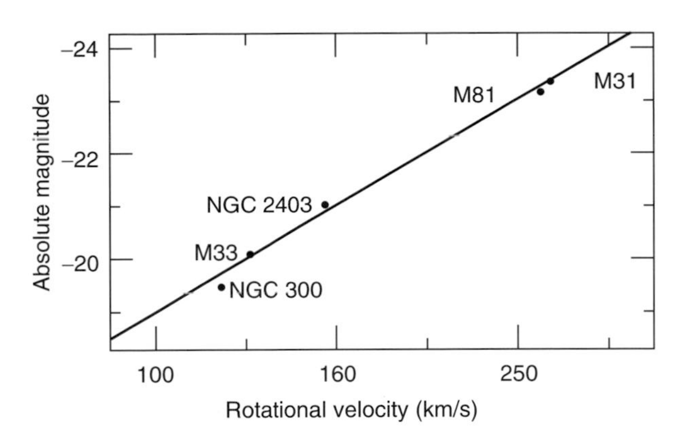

Fig. 1.29. Fisher–Tully relation (see text)

galaxy can be "disky" (high value) or "boxy" (low value). For elliptical galaxies (the same applies for bulges of spiral galaxies) there is an analog to the Fisher–Tully relation, called Faber–Jackson relation. It states that the central velocity dispersion is correlated with the absolute magnitude of the galaxy. Thus we find the more luminous a galaxy, the higher its velocity dispersion. Both relations reflect that the amount of visible matter correlates with dynamical behavior of the galaxy.

Population

From the Milky Way we are already familiar with the concept of populations. The major populations I and II can also be found in other galaxies. Starting with spiral galaxies, their young, massive stars, representing the population I, are primarily concentrated in the spiral arms. The dominant spectral types O and B make them look blue. The old stars of population II, with spectral types around K, make up the bulge, thus looking yellow to reddish. Consequently the bulge-to-disk ratio determines the proportion between both populations. Along the sequence Sa–Sd the fraction of gas, dust, and young stars increases, whereas population II is pushed to the background.

Unlike the case with spirals, population II dominates elliptical galaxies, as they are pure-bulge systems. A few systems show weak disk-structures, indicated by absorption patterns (dust bands). In these cases a small fraction of population I stars is present.

Irregular galaxies offer the reversed picture, being pure population I systems with a large amount of gas. A special case are "blue compact galaxies" (BCD) containing up to 20% gas, which justifies the term *extragalactic HII region.* A violent starburst creates massive stars of types O and B.

Extragalactic Globular Clusters

More than 70 galaxies are currently known to host globular clusters in their halos among them are many large systems [32]. Prominent examples are M 104 or the extreme case of the giant elliptical M 87 in the center of the Virgo Cluster, known to have at least 13,000 globular clusters. Other galaxies, like the edge-on spiral NGC 891 in Andromeda seems to have very few or even none. Astonishingly some dwarf galaxies show globulars, e.g., the nearby Fornax system. Not only the number of globulars varies greatly, but also their distribution. In the case of M 87 and M 49, there are globular clusters lying far outside the halo, making it difficult to decide whether these objects are real globulars or faint background galaxies. Extragalactic globular clusters can be used for distance measurements by assuming an average absolute magnitude of –10 mag for the "brightest" objects in their class. Using only this method a distance of 16.9 Mpc for M 87 was derived.

Most globular clusters are very old, population II objects [33]. But not all, as there are galaxies with violent star formation ("starburst") hosting in their main body a special case of globulars, called *super star clusters* (SSC) or "blue globulars." These are extremely young objects, only a few million to some 100 million years old [252]. Also remarkable is their high luminosity and compactness. They can range up to a hundred times brighter than a "typical" brilliant globular and their absolute magnitude can reach –15 mag. SSCs are associated with huge star forming regions, such as triggered by interaction of their host galaxy with other galaxies or in starburst galaxies.

Cosmic Variety

Similar to stars, the integral quantities of galaxies show a wide range of values (Table 1.5). Dwarf galaxies mark the lower limit. There is a continuous transition from globular clusters to dwarf ellipticals (dE). The galaxies with the lowest mass and luminosity are "dwarf spheroidals" (dSph), containing only a few million stars. Prominent examples are the Local Group members in Fornax and Sculptor (Fig. 1.30). What distinguishes these dwarfs systems from globular clusters? Though the Fornax system is considerably more massive than a typical globular, its stellar concentration is far lower, resulting in an extremely low surface brightness.

At the upper limit are the giant elliptical galaxies, like M 87, or cD galaxies with huge halos. Among the most luminous objects are "ultra-luminous infrared galaxies" (ULIRG), emitting up to $10^{11} L_S$ of infrared radiation. This activity comes from tremendous starbursts, perhaps lasting only 10^7 to 10^8 years, and with star formation rates of 100–1,000 stars per year (Milky Way: 1 star per year). Nearby examples are Arp 220 and NGC 6240 (Fig. 1.31). Not included in the table (listing the "ordinary" cases only) are quasars. They are defined by absolute blue magnitudes of more than –23 mag. The most luminous quasars show $M_B \approx -33$ mag, or over 10,000 times the brightness of the largest galaxies!

According to Sydney van den Bergh, the luminosity is higher for spiral galaxies showing a stronger spiral structure. This leads to the concept of a morphological luminosity classification (in analogy to that defined for stars): class I are "super giants" with $M_B = -21.2$ mag, while class V are "dwarf galaxies" with $M_B = -14.5$ mag. The aim is to calibrate absolute magnitudes of different Hubble types. Here are the classes of some prominent galaxies: the Milky Way I-II, M 31 I-II, M 33 II-III, LMC III-IV, NGC 6822 V.

Table 1.5. Range of integral quantities in comparison with the Milky Way (M_S, L_S = mass, luminosity of the sun)

Quantity	Range	Milky Way
Mass (incl. dark matter)	$10^6 M_S \ldots 10^{12} M_S$ ($10^{14} M_S$)	$1.8 \times 10^{11} M_S$ ($10^{12} M_S$)
Absolute magnitude	–8 mag … –23 mag	–20.5 mag
Luminosity	$10^4 L_S \ldots 10^{11} L_S$	$10^{10} L_S$
Linear diameter	0.1 kpc … 1,000 kpc	30 kpc

Fig. 1.30. The Sculptor dwarf spheroidal system

The "luminosity function" (LF) is defined as the frequency distribution of galaxy luminosities in a fixed volume of space, e.g., a galaxy cluster. At low luminosities, the LF becomes quite uncertain. Nevertheless it shows that dwarf galaxies are most frequent in the universe. Still unknown is the fraction of the recently discovered large *low surface brightness galaxies* (LSB) [34]. The prototype is Malin 1 in Coma Berenices, which lies in the direction of the Virgo Cluster, but is 20 times more distant (Fig. 1.32). LSB-galaxies may contribute to the "missing mass" of the universe [35].

Evolution, Spiral Structure

Hubble assumed his classification scheme to be related with evolution – thus the terms "early" and "late." In this view, elliptical galaxies would evolve to S0- and spiral galaxies; the bulge flattens to a disk. We now believe that the tuning fork must be read inversely. Looking at nearby (evolved) galaxy clusters, the present fraction of elliptical galaxies is high (75%). With increasing distance it becomes as low as 30%, i.e., in the early universe spiral galaxies were much more frequent. This "Butcher–Oemler effect" was confirmed by the *Hubble Space Telescope* (HST). Peering deep into space it finds unmistakable evidence that the universe and its constituents are evolving [36]. In 1995 the HST made a deep image of an isolated $2.3' \times 2.3'$ field in the direction of Ursa Major, the *Hubble Deep Field* (HDF) [37,226]. It shows a large number of young spirals, irregular and interacting galaxies, but only a few ellipticals (Fig. 1.33).

Fig. 1.31. The ultra-luminous infrared galaxy NGC 6240 in Ophiuchus

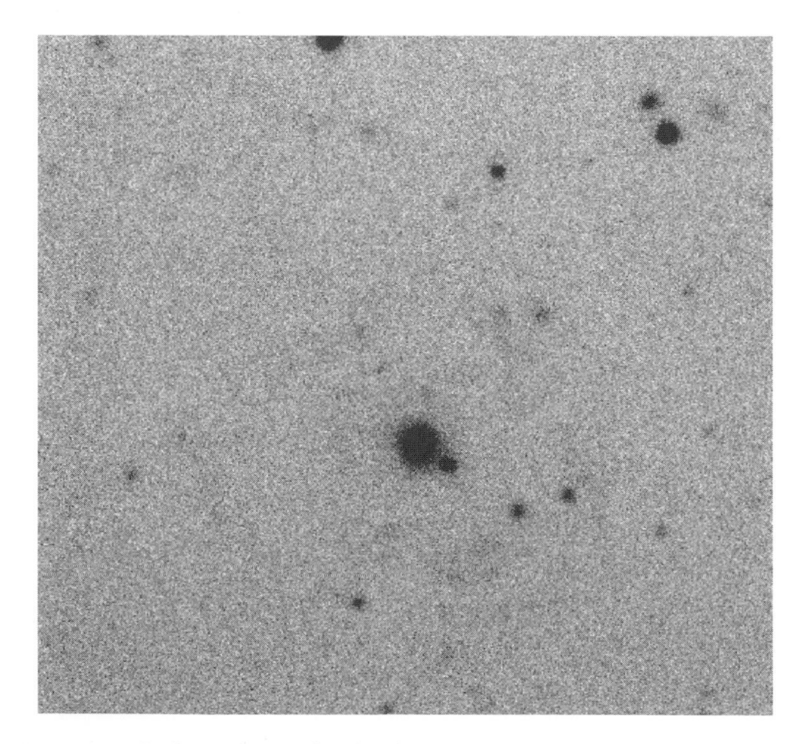

Fig. 1.32. The large "low surface brightness" galaxy Malin 1 in Coma Berenices

Fig. 1.33. Young galaxies in the Hubble Deep Field

How did the first spirals form? They came out of diffuse gas flows or small clumps, falling onto a slowly growing, already rotating disk. But what causes the clumping? It now seems that supermassive, rotating black holes are the obstetricians. But this leads to a classic hen-egg-problem: black holes are the relicts of burnt-out massive stars. Why they can be the seeds for the later galaxies at the same time? Anyway, as they become more compact the rotation speeds up and the structure flattens. The clumpy disk gradually converts into a disk of stars. The young galaxies in the HDF still show the clumpy structure, pointing to high star formation rates. To get a spiral structure, something must disturb the symmetry [1]. In all possible scenarios gravity is the driving force. Perturbing potentials can be massive star forming regions, or – which is probably the dominant – close encounters between galaxies. We may see the effect in M 81 or M 51. The disturbing companion of M 51 is still near (NGC 5195). For M 81 there are actually two candidates, the peculiar systems NGC 3077 and M 82. Probably the encounters are also responsible for the starburst activity in M 82.

Perturbing forces induce "density waves" in the young disk, moving around the center [38]. This motion is independent of the rotation of the disk stars. Places of higher density lead to a compression of the interstellar matter, which triggers stars formation. The newly formed massive stars are like the spray of a water wave. Their high luminosity plus the surrounding HII regions marks the latest position of the density wave – visible as a spiral arm. There is a background of low luminosity stars in the disk, building its "sediment." Between the arms, the density wave has a minimum, thus the activity is low there.

Advancing the density wave, the neighboring zone will be affected. It leaves all objects at their very places (ignoring the rotation of the disk stars for simplification). This is much like a sound wave: the air molecules oscillate around their mean places, while the wave (indicated by the position of maximum density) is moving through the medium. Thus the moving spiral pattern (indicating the position of the density wave maximum) does not contain the same stars. Normally the spiral arms trail relative to the disk rotation, but there could be exceptions of "backward" rotating spirals: NGC 4622 (Fig. 1.34), NGC 3124, or M 64 are candidates. The theory of density waves explains the stability of the spiral structure. If the spiral arms, i.e., all its matter, would move like a whirlpool, they would wind up in a few periods, blurring the pattern.

What's about barred spirals [39]? It seems that the bar phenomenon could be a key to understand spiral structure. Bars are not rare, being present (in its strong SB- or weak

Fig. 1.34. NGC 4622 in Centaurus – a backward rotating spiral?

SAB-form) in 2/3 of all spirals. As young galaxies (seen in the HDF) are mostly "ordinary" spirals, it is likely that a bar is created at a later point in the evolution. It seems to be that a "spontaneous" bar formation (in an isolated galaxy) is not the common way. As nearly all features are related or even due to interaction, so is the bar. An "induced" formation, in an otherwise stable disk, is most likely. Once developed, the bar is a robust feature, remaining over many galactic rotations. Nevertheless, many questions about the formation and evolution of spiral galaxies are still open, and we are faced with rivaling theories [1].

Where are all the early spiral galaxies? They obviously change to elliptical galaxies by gravitational interactions. Model calculations show that the collision of two spirals will likely lead to a boxy elliptical. Disky ellipticals result from the merger of three or more spirals. Such events were most frequent in the dense environment of the early universe. During the collision, the spiral arms of the galaxies are sheared off, forming large tails or plumes produced by tidal forces. The remaining gas and dust is compressed and transformed into stars through a series of starbursts. Thanks to the presence of nearby tidal systems, astronomers can study the details. A prominent example is "The Antennae" NGC 4038/39 (Fig. 1.35). At the end of this interaction hundreds of millions or even billions of years later a combined bulge and halo is left and a young elliptical galaxy has formed. The model calculations fit to the observations: remote clusters or the HDF contain a three times higher number of interacting galaxies and many luminous infrared galaxies, looking much like young ellipticals.

Now back to our home galaxy, the Milky Way. Our galaxy does not seem to be affected by large collisions in the past, but a series of smaller ones. Outside the Milky Way (and

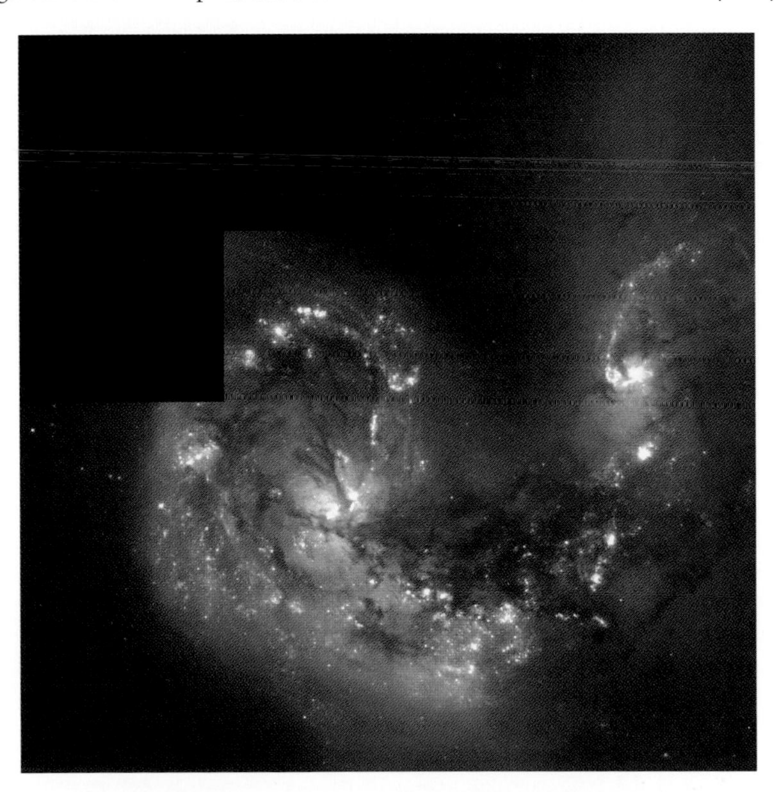

Fig. 1.35. "The Antennae" NGC 4038/39 in Corvus

other LG members) one has detected "high velocity clouds" (HVCs), raining on our galaxy. This is a tremendous reservoir of gas which keeps the star formation process rolling [253]. New data of the Sloan Digital Sky Survey suggest that there is an outer ring of stars circling the Milky Way. This may be the tidal debris of several merged dwarf galaxies. Another speculation concerns the bright star Arcturus, which could be an ancient member of a merged dwarf galaxy population. But these were all minor collisions. The Milky Way will meet his fate in about 5 billion years as it will collide with the M 31, the immense Andromeda Nebula. The nightsky of a hypothetical earth will then be brightened by starburst regions, blue knots, and massive newly formed star clusters, plus a growing number of supernovae. Eventually the familiar band of the Milky Way disappears, replaced by a huge homogenous, spherical star cloud. The new home of the combined civilizations of the Milky Way and Andromeda will be a giant elliptical galaxy!

Quasars

Quasars are young objects, in two different meanings: they were discovered quite recently (in the 1960s) and are cosmologically young objects – in their proper time frame. Thanks to their remoteness we can still observe them – and we need only a small telescope!

Let's begin with a short note on distance of galaxies when calculated by means of the redshift z. The result depends strongly on the cosmological model and the definition of "distance," which is not an obvious quantity in general relativistic cosmology [40]. We may talk about the "light travel time" [41], which gives a distance by multiplying with the speed of light (c). This value is both different from the distances at emission and reception of the light, respectively. The latter is an "instantaneous" distance, which is much larger due to the ongoing expansion. One such measure is the "proper motion distance" (d_M), which we choose here (calculated with $H_0 = 71$ (km/s)/Mpc). Other distances, like the "luminosity distance" (d_L) or the "angular diameter distance" (d_A), can be derived through $d_L = d_M \cdot (1 + z)$ and $d_A = d_M/(1 + z)$.

The first quasars were noticed when optically identifying strong radio sources. The image appears stellar ("quasi stellar radio source") and are often blue. Therefore it is not astonishing that many quasars were already cataloged as "faint blue stars." The object HZ 46, from the list of Humason and Zwicky is a remarkable example. It was discovered in 1947, but turned out to be the first compact extragalactic object. But it fails to be a quasar, being not luminous enough.

Quasars show an isotropic distribution on the sphere, a clear sign of their extragalactic nature (galactic objects are concentrated toward the band of the Milky Way). That and their extremely high redshifts provide the needed proof they are "extragalactic" objects. Many quasars also show the presence of strong emission lines (and absorption lines too, due to the intergalactic medium [42]).

We now know that quasars are extremely luminous nuclei of galaxies [2,43]. The activity of a quasar originates from an extremely small volume. The proposed engine is a supermassive black hole lying at the core. The attracted matter cannot stream in radially, but is forced to approach the black hole in spiral orbits. This rotating structure is called an "accretion disk," and lies just outside the event horizon. Due to friction the matter heats up to very high temperatures, producing strong emission lines in the quasar spectrum. The black hole also shows an extremely strong magnetic field, forcing charged particles to fall in at the poles, perpendicular to the accretion disk. Acceleration induces synchrotron radiation, focused into jets, which are emitted at both poles. They interact with the intergalactic medium at large distances. Thus many quasars are strong radio sources, displaying huge lobes of ionized matter and energy. There are some examples, where jets can be seen optically, e.g., 3C 273 in Virgo.

Many quasars show a faint halo, which is the "host galaxy" (Fig. 1.36). The original definition of a quasar as a "quasistellar" object is now less significant. The transition from stellar to galaxy-like objects, e.g., AGN is smooth. BL Lacertae objects are a special case of quasars showing synchrotron radiation, but practically no thermal emission. Thus, lacking emission lines, it was very difficult to determine their redshift. The measurement is only possible, if a host galaxy can be detected. By cutting out the central source with a tiny diaphragm it is possible to measure the spectral lines of the galaxy. We now know that a BL Lacertae object (or "blazar") is a quasar, where the jet is pointed toward us, outshining the accretion disk. Which type of quasar we see therefore depends on the spatial orientation. Often the term "quasar" is used for all types, but it is occasionally necessary to distinguish.

Beside all new types and definitions, one fact about quasars still stands: there is nothing comparable in the universe having such a huge, steady, and isotropic output of energy. What's about "gamma ray bursts" (GRB)? It's true that they show an even higher luminosity. Some GRBs have been proposed to be a type of *hypernova* and are thus transient, stellar events. In GRBs it is theorized that the observed burst of energy is focused into extremely tight jets that are pointed toward us by chance [223].

Due to their huge luminosity, quasars can be detected at distances of 10 billion ly or more. This opens a view into the earliest times of our universe, in which most galaxies

Fig. 1.36. Quasar PKS 2349-014 in Pisces and its host galaxy, the dominant member of a remote galaxy cluster

(and quasars too) were young objects, located in dense clouds of matter. Thus it is not astonishing that many quasars were found to be dominant members of remote galaxy clusters (Fig. 1.36). This leads to an understanding of their formation. We have already discussed that young galaxies in the early cosmos were often subjected to collisions. The merged nuclei formed supermassive black holes. With a substantial supply of interstellar matter there was enough "food" to feed the "monster" (the massive black hole), leading to an extreme nuclear activity – or a quasar. The quasar phenomenon seems to be a quite ordinary phase in the early history of galaxies.

Obviously quasars must die out at some time. Permanent collisions, which trigger an extremely high star formation rate, eventually thin out the interstellar matter. The black hole runs out of "food" and the infall becomes unsteady. If a larger flux of matter becomes available, it produces an optical burst. Thus many of these objects are variable. The intermediate phase toward a quiet nucleus is called AGN ("active galactic nuclei"). Such objects, like Seyfert Type I galaxies, are still quasar-like but less luminous. The optical criterion is the absolute magnitude: quasars are defined to be brighter than $M_B = -23$ mag. For BL Lacertae objects such a limit cannot be given, as their brightness does not result from thermal radiation. They are defined through their typical (continuous) synchrotron spectrum. This is the reason why some bright galaxies, e.g., NGC 1275 are occasionally classified as blazars.

The lack of nearby quasars shows that at present the chance for creation is pretty low. Nevertheless, the ultra-luminous infrared sources Arp 220 and NGC 6240 maybe new quasars in the making (Fig. 1.37 and Fig. 1.31). It fits into the picture as that NGC 6240 bears two massive central black holes lying only 3,000 ly apart.

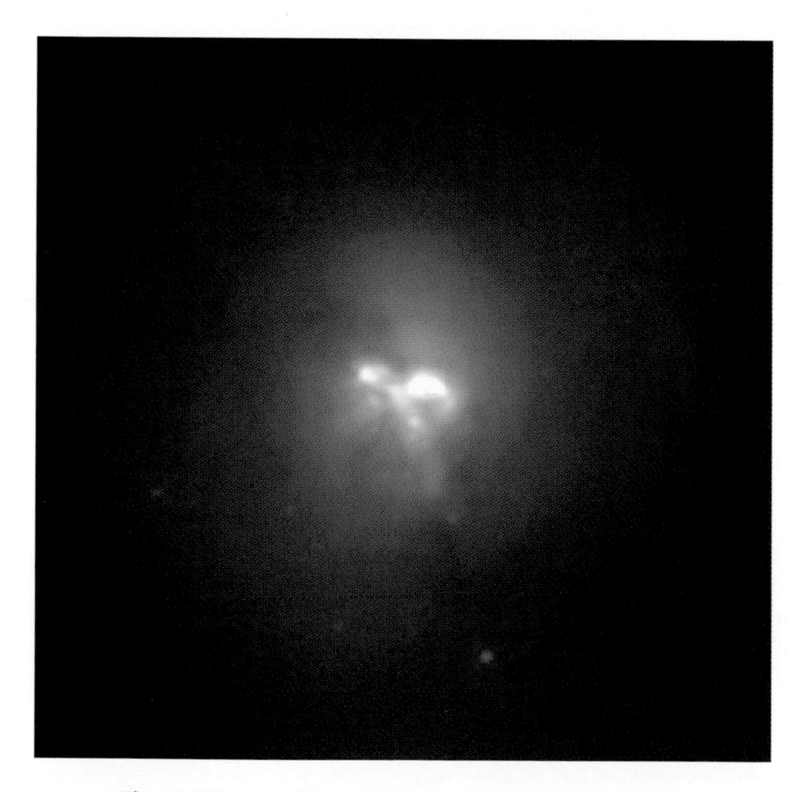

Fig. 1.37. Arp 220 in Serpens – a quasar in the making

Chapter 2

Pairs, Groups, and Clusters of Galaxies

Isolated galaxies in space are rare. Galaxies generally tend to form pairs, multiple systems, and groups ranging from a dozen members, to huge rich clusters, hosting thousands [44]. This behavior is well known from the stars forming our own Milky Way. Galaxies and clusters of galaxies belong to the largest structures in the universe, and their observation will lead the amateur into deep space and simultaneously as a view back into a remote past. The light of the most distant objects travels many hundred million or even billions of years to reach our eye. Galaxy clusters, especially those dominated by extremely luminous cD galaxies, are easily observable and in many respects attractive targets for amateur telescopes.

Let us first discuss how clusters of galaxies are placed in the hierarchy of the universe. This is strongly related with the question of their formation and evolution. After cataloging a large number of clusters, different cluster types came to light, and it is interesting to look at the various classification schemes. As for individual galaxies, morphology and structure are essential for the study of the evolution and dynamics. But clusters are not mere agglomerations of finished galaxies, they "act back" on their members, forcing them to change their structure. Therefore, as was already pointed out in chapter 1, the evolution of galaxies and clusters is strongly related.

Galaxies and Clusters in the Hierarchy of the Universe

Local Group and Local Supercluster

Based on the study of the nearest galaxies, the Milky Way is part of a small cluster, called the "Local Group" (LG) [45]. Our galaxy, plus our neighbors the Andromeda Nebula (M 31) and the nearby Triangulum Nebula (M 33) are the dominant members. The LG contains at least 38 systems, mostly irregular, elliptical, or spheroidal dwarf galaxies in a volume of 2.4 Mpc in diameter (Fig. 2.1). Best-known "minor members" are the Large and Small Magellanic Clouds (LMC, SMC), both companions of the Milky Way. Two other prominent dwarfs are associated with the Milky Way too; these are the Fornax and Sculptor systems. The nearest member is the newly found dwarf elliptical galaxy in Canis Major [46], just followed by the "Sagittarius Dwarf Elliptical Galaxy" (SagDEG), located on the opposite side of the Milky Way, hidden by the bulge [7,47]. Freely visible, it would

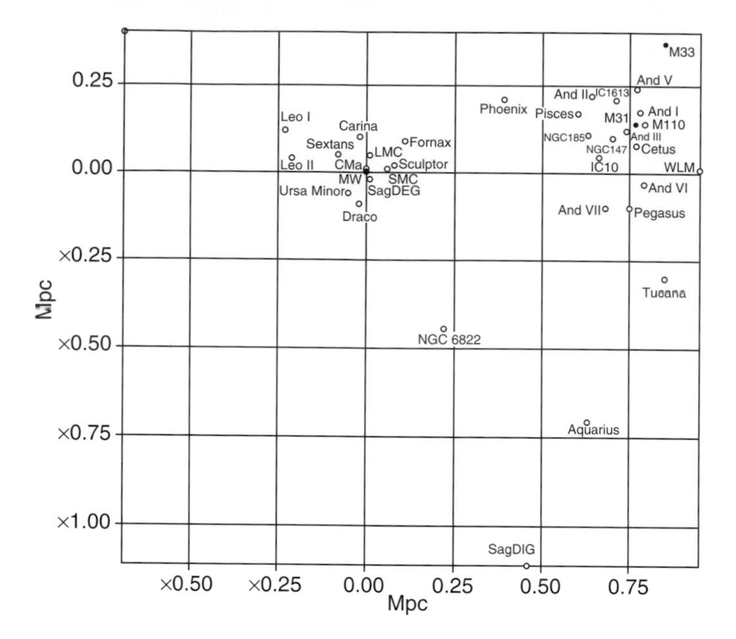

Fig. 2.1. The Local Group (main galaxies plotted)

shine at an integrated magnitude of $V = 3.6$ mag. The Andromeda Nebula has a number of companions too: the largest are M 32 and NGC 205, while the smaller systems are named And I–X.

Using both direct distance measurements and Hubble's law it is clearly evident that groups and clusters are not random concentrations on the sphere, but in fact three-dimensional dynamical structures. Groups resembling our Local Group are well studied. Examples are the UMa group around M 81/M 82 or the CVn group, centered on M 51, at distances of 3–4.5 Mpc. But these are only outer condensations of a much larger collection known as the Virgo Cluster, about 20 Mpc distant. The "core" of the cluster lies near the giant elliptical galaxy M 87 and nearby M 84/M 86. This cluster is so massive that all surrounding groups, including the Local Group, are affected by its gravitational pull. This effect is commonly called the "Virgo flow." The gravitational force disturbs the smooth Hubble expansion, thus making it difficult to determine the present Hubble parameter (H_0). That's the reason why distances derived from Hubble's law are not reliable on "small" scales (low z).

In an early plot of the 1,246 brightest galaxies from the Shapley–Ames catalogue a striking asymmetry between the northern and southern galactic hemispheres is visible (Fig. 2.2). The excess of northern galaxies is due to the "Local Supercluster." It defines the next step in the cosmic hierarchy, actually a cluster of clusters. In its center the Virgo Cluster dominates with some 3,000 galaxies, surrounded by galaxy groups of various sizes, one of them is our Local Group. The Local Supercluster lies just "above" (north of) the Milky Way and this explains the asymmetry between the two hemispheres. Consequently the Local Group is located at the edge of the Local Supercluster – being only an appendix of the much larger structure.

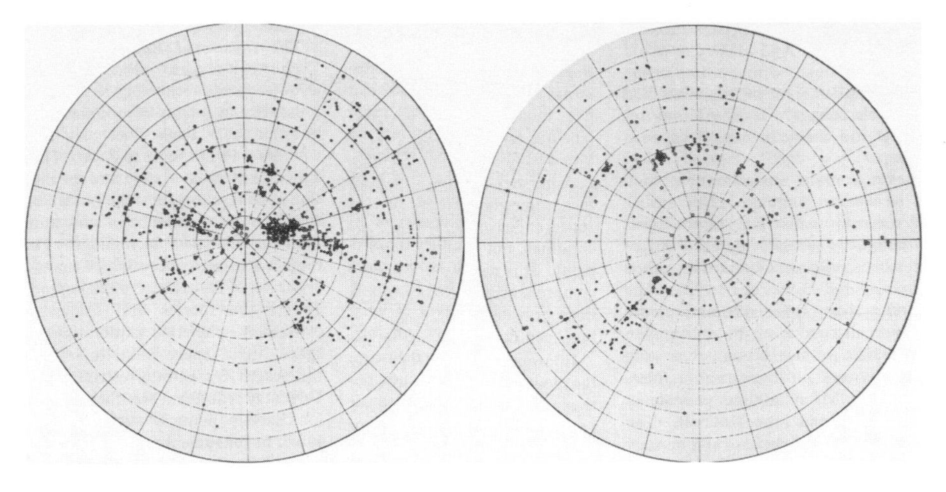

Fig. 2.2. Galaxy distribution in the galactic hemispheres, based on the Shapley–Ames data. The northern hemisphere (left) is dominated by the Local Supercluster

Distant Clusters of Galaxies and Redshift Surveys

At larger redshift more distant galaxy clusters appear, like those in Perseus, Coma Berenices, Hercules, Corona Borealis, and Hydra [48]. With ever-greater distance, clusters and its members should look smaller and fainter. By the magnitudes of the dominant cluster members, assuming to have comparable luminosities, one can estimate the distance. This relationship can be used to extend the Hubble law into deep space.

By popular convention, the "classic" galaxy clusters (e.g., Coma, Perseus, Hercules) which all lie at distances of at least 100 Mpc, have been integrated in respective superclusters of the same name. These superclusters are even bigger than the Local Supercluster and belong to the largest known structures in the universe.

The modern picture of the universe is essentially based on massive galaxy surveys. To paint the cosmic tapestry, and for a better three-dimensional view, astronomers have relied on large redshift surveys with thousands of galaxies [49]. The Harvard Center for Astrophysics (CfA) survey by Huchra and Geller contains 24,000 galaxies and presents spatial slices of the cosmic structure [50]. The large-scale arrangement of matter looks like a Swiss cheese or foam. Most galaxies are concentrated in superclusters, connected by long filamental structures (a prominent example is the "Great Wall"), made of smaller clusters and groups. Moving away from these structures, the space is quite empty. Giant voids, containing only a few galaxies, can have diameters of 50 Mpc or more, so the cheese is pretty holey. This structure discovered by Huchra and Geller was confirmed out too much greater distances by other massive campaigns. Examples include the 2dF redshift survey [51], made with the "2-degree field system" at the Anglo American Telescope (Fig. 2.3), the 2MRS (*2MASS Redshift Survey*), both with 250,000 galaxies, and the redshift survey based on the *Sloan Digital Sky Survey* (SDSS), containing one million galaxies [52,231]. At present we overlook a volume of nearly 1,000 Mpc and we believe that the foam will extend to even greater distances and that no new structures ("super-superclusters") appear on the cosmic scene.

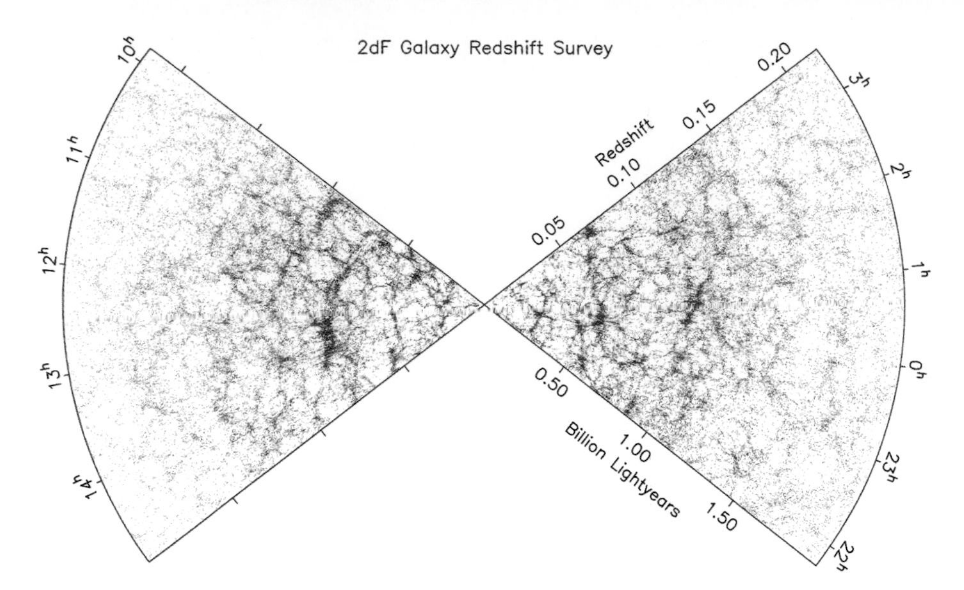

Fig. 2.3. Slice of the local universe up to 2 billion ly from the 2dF redshift survey (the giant arc in the left slice between 0.5 and 1.0 billion ly is the "Sloan Great Wall")

Evolution of Large Scale Structures

Nature's finite speed of light is a useful gift, otherwise cosmology might be much more difficult. It permits us to look into the past, revealing the evolution of the hierarchical structure and its constituents. What about the case of a hypothetical infinite speed of light? We would be faced with an instantaneous cosmic situation and must derive the evolution of objects, being in different parts of their lives from the observable variety. Much like Milky Way stars, where light travel time is small against their life spans.

Receiving information about different evolutionary stages, two opposite models were developed: the hierarchical "bottom-up" model, mainly supported by Peebles, and the "top-down" model, developed by Zeldovich. The simple alternative is: were the smaller objects (galaxies) first to form the larger structures (clusters, superclusters) or vice versa. In the Zeldovich scenario huge, flat clouds of matter ("pancakes") condensate to smaller parts to build "protogalaxies." Thus we are faced with aggregation ("bottom-up") vs. fragmentation ("top-down").

The first efforts to simulate the evolution of cosmic structures based on gravitational instabilities in the primordial matter were not very successful. The Universe was not old enough to create the observed patterns. But these calculations left out the influence of dark matter. All went well with the aid of additional, unseen masses, accelerating the evolution. Recent computer simulations of millions of bodies under the influence of "cold dark matter" (CDM) match the observed structures pretty well (Fig. 2.4). CDM is most probably made of "cold" (i.e., slowly moving) elementary particles, which are still undiscovered due to their extremely weak interaction with ordinary matter. Any known particles like neutrinos, known to have small masses and moving nearly with the speed of light (thus "hot dark mater") are out.

The CDM model confirms the "hierarchical" evolution of structures. The observed inhomogeneous distribution of matter primarily results from density fluctuations in the primordial soup, which is graved as tiny deviations in the temperature field of the 3 K

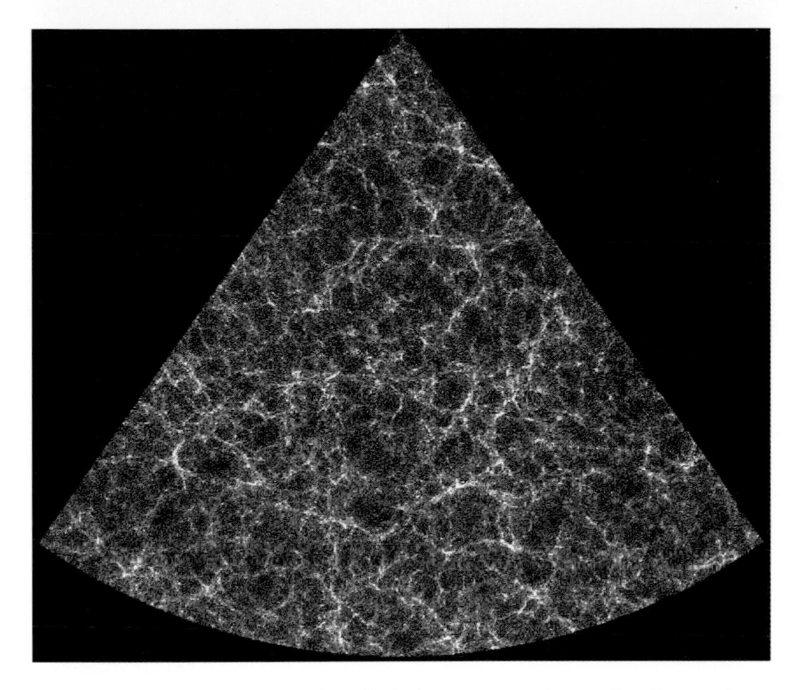

Fig. 2.4. Computer simulation with cold dark matter producing the observed structures

background radiation. These are the very seeds of the cosmic structures. The first inhomogenities in the CDM distribution aggregate ordinary matter to form protogalaxies in a pretty short time. Gravitational attraction keeps the galaxies together to accumulate clusters.

Classification and Dynamics

After outlining the cosmic hierarchy and the building of structures, let's take a look at the different morphological types of galaxy agglomerations, their classification, and dynamics. It is quite natural to sort things by increasing number of members: from pairs, multiple systems, groups, small clusters, rich clusters to superclusters.

Pairs of Galaxies

Pairs of galaxies are like double stars; they can be mere "optical" (chance alignment) or "physical" (gravitationally bound). One of the most striking examples of an optical pair is NGC 3314 in Hydra, where two spirals are superimposed (Fig. 2.5). Detecting galaxies a few arc minutes apart, it is difficult to decide, which case is present. If there is a large brightness difference, a physical connection looks unlikely, but nevertheless cannot be ruled out. It is possible that an ordinary galaxy has a nearby dwarf companion (think about observing our Milky Way plus LMC from a distant). We therefore have to register the relevant galaxy types. In a pair of spirals in which one is much fainter, this is obviously a background object.

Fig. 2.5. The stunning optical pair NGC 3314 in Hydra

There are two possibilities to verify a physical pair. The first, qualitative way is to look for tidal effects. The early studies of double and multiple galaxies used this very criterion. Members often show tidal distortion, resulting from close encounters or even collisions [53]. Their shape depends on the objects' orbital paths. Frontal collisions may form ring galaxies. In other cases large streamers, tails, bridges or jet like features are created. The dynamics of such processes can be visualized through computer simulations, e.g., in the cases of M 51/NGC 5195 or the Antennae NGC 4038/39. Even "at home" there is interaction: the Milky Way disturbs the LMC, producing a tail called the "Magellanic Stream" containing six hydrogen clouds. It is now evident that any contact triggers star formation, which can be detected by strong infrared emission.

The quantitative way to verify a physical connection of galaxies is measuring their distances. Physical pairs must show comparable redshifts. But this has led to some controversy, as shown best in the extreme case of "galaxy–quasar pairs."

Galaxy–Quasar Pairs, Gravitational Lensed Quasars, and Real Double Quasars

The classic example is NGC 4319 and Mrk 205 (Fig. 2.6). Mrk 205 (V = 15.2 mag) is located only 42" south of the SBa-galaxy NGC 4319 (V = 11.9 mag), embedded in its diffuse halo. Actually the object is more an AGN than a quasar. It shows a variability of 0.5 mag and can reach the quasar definition level of $M_B = -23$ mag. Could this be a chance alignment or a true physical pair? The advocate for the latter is Halton Arp [54]. The problem is that Mrk 205 shows a 14 times higher redshift ($z = 0.070$), calculating a spatial

Fig. 2.6. The optical galaxy–quasar pair NGC 4319 and Mrk 205 in Draco

separation of 265 Mpc. But Arp believes that there is a "bridge of light" between both objects [241]. He interprets the redshift difference as a kinematical (instead of a cosmological) effect: the quasar must be ejected from the galaxy with very high speed. This is in conflict with all other observations and some major questions remain: How to explain the redshift in general? Why there are no cases where objects are ejected toward us, showing a blueshift? No one, except Arp, is willing to give up the cosmological interpretation of redshifts. We are now sure that Mrk 205/NGC 4319 is a chance alignment, as the "bridge" was a type of image processing artifact. Another case of an optical pair is NGC 3067 ($V = 12.1$ mag) and 3C 232 ($V = 15.8$ mag) in Leo Minor. The quasar ($M_B = -26.7$ mag, distance 4,755 Mpc) is located 2′ north of the SBab-galaxy.

Are their pairs of quasars? Quasars are extremely luminous nuclei of young galaxies. Like double galaxies, we therefore expect real double quasars to exist. Unfortunately, there is a special case, which looks like a real pair. But this is due to gravitational lensing. According to Einstein's General Relativity [55], the light of a remote object can be deflected by a large mass (galaxy, cluster of galaxies), lying in the line of sight. When light of a remote quasar passes a galaxy, it can split into two or more images, separated within a few arcseconds [56]. Their spectra are nearly identical, which is the very criterion of a gravitational lens. In case of variability, all "components" show synchronous brightness variations.

The first example, the "Double Quasar" in Ursa Major, Q 0957+561, was detected in 1979. It is located 14′ north of the bright galaxy NGC 3079. The images have magnitudes of $V = 16.7$ mag and $V = 17.0$ mag, respectively, and are separated by 6.2″. Note that it is not correct to call an object like the "Double Quasar" a "gravitational lens" (like e.g., in

[235]). The lens is the massive foreground object bending the light rays of the quasar (in this case a galaxy). Thus one must speak of a "gravitational lensed quasar."

At present around 50 cases are known (a sample of 19 objects is presented in [246]). Prominent examples of multiple images are the "Triple Quasar" PG 1115+080 in Leo (combined 15.8 mag, separations 2.1″ and 2.7″), the "Einstein Cross" Q 2237+0305 (V = 16.8 mag; Fig. 2.7) in Pegasus, and the "Clover Leaf" H 1413+117 in Bootes (V = 17.0 mag). A spectacular case is APM 08279+5255 in Lynx with a redshift of z = 3.87 that places the quasar among the most distant objects to be seen visually (V = 16.5 mag). Due to its brightness and distance, it was first thought to be the "most luminous object" in the universe; with M_B = −32.2 mag it would be as bright as the sun at a distance of 350 ly. But a detailed analysis showed two lensed images of equal magnitude at a 0.4″ separation. Gravitational lensing amplifies the light, so the true luminosity of the object is actually lower. The current record holder is MG2 J165543+1949 in Hercules with M_B = −31.4 mag (V = 16.2 mag, z = 3.26).

What's about real double quasars? It is surprising that the more exotic case of gravitational lensed objects was first discovered, and one had to wait another 10 years to find the first physical pair, OM-076 (Fig. 2.8). In this case the components show different spectra (and not identical copies) at the same redshift. At present more than 10 pairs are known. CT 344 in Sculptor shows the closest separation of 0.3″. The brightest case is HS 1216+5032 in Coma Berenices with B = 17.2 mag and B = 18.6 mag (separation 8.9″). There is another characteristic, which differs from gravitational lenses – the degree of variability. Keeping in mind that the brightness changes with the amount of matter the central black hole is being fed, we can expect different, not synchronous variations for both objects.

Fig. 2.7. The "Einstein Cross" in Pegasus with its 15 mag lensing galaxy CGCG 378-15

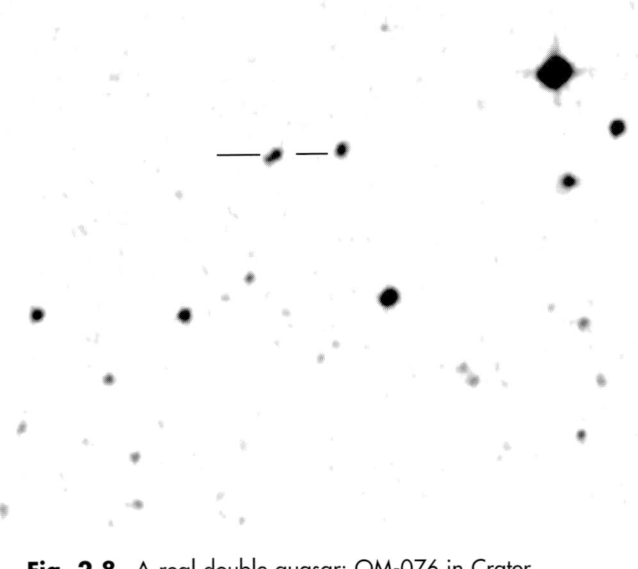

Fig. 2.8. A real double quasar: OM-076 in Crater

From Galaxy Groups to Poor Clusters

Going back to "ordinary" galaxies we can find different degrees of organization. The first step in the hierarchy are multiple systems comprised of a few members and groups, as defined by a population of up to 20 galaxies. Such structures show a broad range of densities, from loose to compact, and different shapes, from spherical structures to long chains. One of the most prominent chains is located in the heart of the Virgo Cluster: "Markarian's Chain," stretching from M 84 and M 86 through to NGC 4477 [224]. An example of an extremely compact group is Shkh 1 (Fig. 2.9), first thought to be a remote globular cluster of old red stars. But the redshift shows something very different: it is a group of 20 compact elliptical galaxies covering an area of only 1.5′. They may be the results of merged spirals, loosing most of their interstellar matter by repeated encounters in the dense environment.

In a stable group, gravity and internal motions must balance. The observed redshifts scatter around a mean value. The amount of variation is measured by the "velocity dispersion" (already known from random motions of stars in elliptical galaxies). It is typically around 250 km/s. There are cases in which one galaxy shows a discordant redshift, indicating a much lower or higher radial velocity, than the mean. One must critically investigate, if this behavior is dynamical ("ejection"), or due to a chance alignment. A controversial case was NGC 7320 in Stephan's Quintet in Pegasus, which shows a considerably lower redshift than the other four members [229]. This is actually due to a much closer distance and thus a foreground object.

Fig. 2.9. Shkh 1, a compact group of compact galaxies in Ursa Major

Small clusters, like our Local Group, contain 20–100 members. Often the term "poor cluster" is used to distinguish from "rich cluster." Many have a dominant galaxy, a giant elliptical or even one of type cD [57], though not always centrally located. We have already mentioned that such galaxies are extremely massive and luminous, surrounded by a dense halo up to a million light-years in diameter. Examples are NGC 1129 or NGC 6051. The velocity dispersion in poor clusters indicates that galaxy collisions play a fundamental role. Through computer simulations it is evident that the creation of cD galaxies is a quite ordinary phenomenon and more than half of them end with multiple cores due to cannibalism. The dense halos result from the accumulation of tidally stripped matter.

Rich Clusters

Rich clusters are those whose membership can include thousands of galaxies. The standard catalog was compiled by George Abell using the *Palomar Observatory Sky Survey* (POSS). His work on *The Distribution of Rich Clusters of Galaxies* presents a sample of 2,712 clusters [58], denoted by "A" (sometimes "ARC" or "AGC" is used). The selection is based on three criteria: distance, richness, and compactness [59]. Since Abell did not measure the redshift, "distance" is defined by the photographic magnitude of the 10th brightest cluster member. Using the value of the third brightest member, "richness" is defined (Table 2.1). The "compactness" is derived from the number of members within the cluster radius, for which the "distance" is taken into account.

Abell was well aware that his catalog is incomplete at the low richness end and he defined a homogeneous subsample of 1,682 clusters, which is proved to be 85% complete. He classified clusters as "regular" or "irregular." Regular clusters are roughly spherical with a strong concentration toward the center. Irregular clusters, like the Hercules Cluster (Fig. 2.10), are not symmetric or concentrated. The Abell scheme also includes an intermediate type. An example is A 194, centered on the Pisces group with NGC 541 as the dominant galaxy.

There is a remarkable morphology–density relation, known as the "Butcher–Oemler effect." Regular clusters, like the Coma Cluster (A 1656), contain 70–80% galaxies of type E or S0. Irregular clusters, like the Hercules Cluster (A 2151), show all types of galaxies,

Table 2.1. Definition of distance and richness class for Abell clusters

Distance class (*D*)	Magnitude of the 10th brightest member
1	13.3–14.0
2	14.0–14.8
3	14.9–15.6
4	15.7–16.4
5	16.5–17.2
6	17.3–18.0
7	Fainter than 18.0
Richness class (*R*)	**Number of galaxies in the interval m_3 to m_3+2**
1	30–49
2	50–79
3	80–129
4	130–199
5	200–299
6	300 or more

Fig. 2.10. The Hercules Cluster A 2151 as an example of an irregular galaxy cluster

50% are spirals. This can be explained by evolution. We can see clusters at different ages while looking back in time. The observations indicate that remote irregular clusters contain an excess of blue spiral galaxies, while nearby regular clusters show more elliptical galaxies. The blue spirals undergo bursts of star formation, triggered by collisions. Such objects are frequent in the HDF. As there are no examples in nearby clusters, galaxy populations in clusters obviously have evolved significantly over the past few billion years [60].

Thus we get the following scenario: spiral galaxies came first, forming irregular clusters. In the intermediate phase a considerable fraction of the spirals merge to convert into elliptical galaxies. This transformation results in a regular cluster, showing the present distribution of galaxy types. In rich clusters ellipticals always win the race. This is not true for field galaxies. In less crowded regions of space, the survivability of spirals is much higher. Such is the case in our neighborhood, as all of the large galaxies are spirals. In order to find the nearest (non-obscured) "normal" elliptical galaxy, NGC 3376, we must travel 13 Mpc.

Abell's catalog can be used to determine the cluster distribution and the "luminosity function" (LF), e.g., the number of galaxies per magnitude (or per magnitude interval for the *differential* LF). The Virgo Cluster – which fails to meet Abell's criteria for a rich cluster and thus is not in his catalog – is also an irregular. It is near enough to show many dwarf galaxies (at the faint end of the LF) and is less clustered than the larger types. Dwarfs are rare at the cluster center, being easily merged there.

What about the 9,133 clusters listed by Fritz Zwicky in his *Catalogue of Galaxies and of Clusters of Galaxies* (CGCG)? They are characterized by density (open, medium compact, and compact), population, diameter, and a distance estimate (through the brightness of its members). No redshift information is given. For "open" clusters it is sometimes difficult to comprehend the accumulation and many cases might not be real. Since Zwicky's cluster criteria are weaker than Abell's, the catalog was of no large importance. An essential deficit is that due to its definition the cluster size appears distance dependent.

Classification of Galaxy Clusters

Abell's rough classification is extended by the schemes developed by Bautz & Morgan (BM) and Rood & Sastry (RS). The main criteria are concentration, shape, and the galaxy types involved. If a cD galaxy is present, the cluster is of type cD (RS) or class I (BM), respectively. The BM classification is quite simple as there are only two more classes: type III clusters host no outstanding galaxies and II is intermediate. The BM-scheme continues with classes B, C, L, and F, depending on the number of first ranked galaxies (mostly ellipticals). In class B a pair dominates (e.g., in the Coma Cluster), C contains a core of 3–4 bright galaxies, L describes a linear arrangement of the brightest members (e.g., in the Perseus Cluster, Fig. 2.11) and class F assigns a flattened distribution of the 10 first ranked galaxies. The SM classification ends with I for irregular clusters, where all different types occur with comparable magnitudes (e.g., in the Hercules Cluster). A common structural feature of galaxy clusters is the occurrence of subclustering, which gives a lumpy cluster appearance. In the evolutionary (hierarchical) picture, clusters are formed by the progressive fusion of an inhomogeneous system of subclusters.

Some extremely remote clusters of galaxies can only be noticed by inspecting deep images of the area around giant radio galaxies. This again confirms the concept of massive central galaxies in clusters. Examples are the clusters around Cygnus A, 3C 66A or 3C 295. All were first discovered and cataloged as strong radio sources. While 3C 66A is a quasar, 3C 295 and Cygnus A were classified as giant elliptical galaxies. But new observations show that Cygnus A hosts an active quasar core hidden by a gigantic doughnut-like

Fig. 2.11. The Perseus Cluster A 426 is pretty elongated, containing a large number of E and S0 galaxies

dust ring (Fig. 2.12). Cygnus A, lying at a distance of 266 Mpc ($z = 0.065$), is 1,000 times more massive than the Milky Way.

Missing Mass and Cosmic Effects

Most galaxy clusters range in size from 3 to 10 Mpc. Their masses are typically around 10^{15} M_S. They can be derived using the "virial theorem," expressing the balance between gravity and velocity dispersion in a stable agglomeration (the dispersion is typically around 750 km/s). According to Zwicky, the virial mass is 10 times higher than the luminous mass, as indicated by the visible galaxies. Clusters are thus much more massive than they look. We're already familiar with the "missing mass" problem from the rotation of galaxies. For galaxy clusters a much larger amount of "dark matter" is needed to close the gap. The true physical nature of this matter still remains a mystery. A certain portion of this matter may be located in the halos of the cluster members. Satellites have detected a hot intracluster gas, emitting X-rays (Fig. 2.13) in a number of clusters. However, its origin is still unknown. This "hot" halo surely contributes to the missing mass, but actually also intensifies the problem as a large amount of additional mass is now needed to confine the turbulent gas in the cluster!

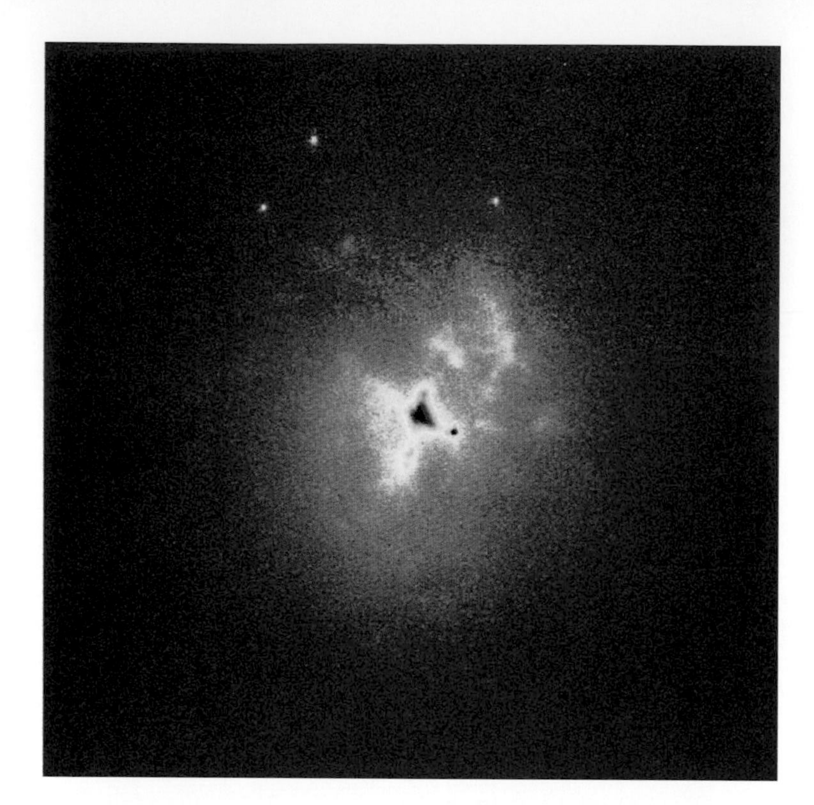

Fig. 2.12. Cygnus A, a giant "dusty" elliptical galaxy with a quasar core

Rich clusters produce interesting observable effects useful to determine the cosmological quantities, thus being steps in the cosmological distance ladder. Due to their tremendous mass, they can act as gravitational lenses, distorting the light of more distant galaxies into arcs (Fig. 2.14). Arcs indicate the distribution of matter in the cluster, presenting an alternative way to determine the cluster mass. The result is compatible with the calculated virial mass. Clusters also influence the cosmic background radiation due to interaction of their photons with the electrons of the hot intracluster gas (Zeldovich–Sunyaev effect). This lowers the brightness temperature of the 3 K cosmic background radiation as seen through the cluster center by around 1 mK and weakens their intensity. Both effects can be used to determine the Hubble parameter – independent of any kind of distance estimate. This leads to value of H_0 compatible with 71 (km/s)/Mpc.

Superclusters

We observe a large scale second-order clustering, which are actually superclusters beyond the Local Supercluster. They are identified through calculating cluster correlations in the Abell sample by statistical methods. This is much like using high mountain peaks trace mountain chains. Superclusters, hosting up to 15 clusters have typical diameters of 100 Mpc as in the case of the Coma Supercluster. As peculiar motions in superclusters are not yet known, the virial theorem is not applicable. We therefore can only add the masses of the constituents, resulting in supercluster masses around $10^{16}M_S$.

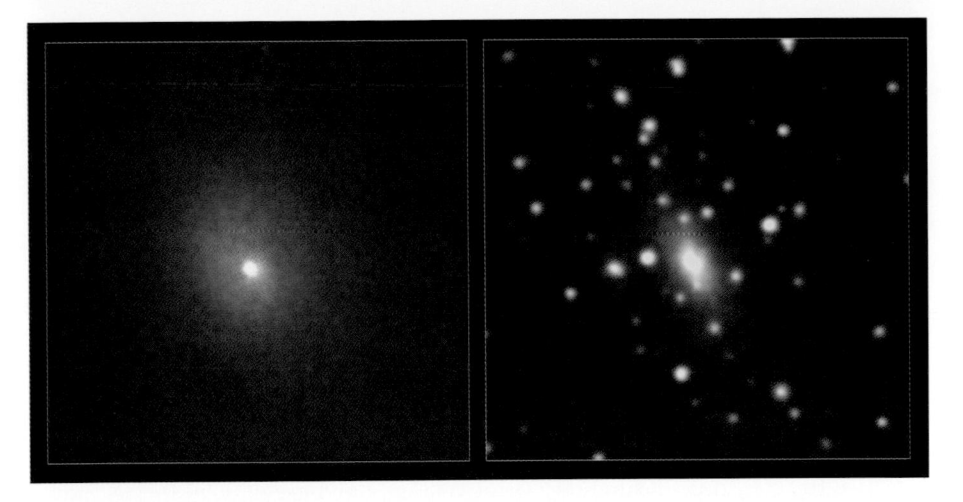

Fig. 2.13. X-ray emission from the rich cluster A 2029 in Serpens with central cD galaxy IC 1101

Fig. 2.14. Arcs of lensed background galaxies in the rich cluster of galaxies A 1689 in Virgo

One of the most massive supercluster aggregates is the "Great Attractor" (GA), having a mass of $5\times10^{16}M_S$. Its gravitational attraction on the Local Supercluster slows down the "local" expansion rate by 500 km/s. The GA is located behind the Hydra-Centaurus Supercluster at a distance of 200 Mpc. Unfortunately, this is in the direction of the Milky Way's "zone of avoidance," which absorbs most of the distant light [7]. Optically the center of the GA is marked by the rich cluster A 3627, contributing 10% of the total mass.

Some superclusters show a distinctly flattened shape. Examples are the Local Supercluster, with a diameter of 50 Mpc or the Coma Supercluster, which is almost double in size. This shape is probably a relict of a "pancake" distribution (the Coma Supercluster was thus characterized as a "Zeldovich disk"). Others are like filaments or chains, e.g., the Perseus-Pisces Supercluster or the "Great Wall." This is a massive structure of $2\times10^{16}M_S$ extending 340 Mpc × 120 Mpc and with a thickness of only 10 Mpc lies about 130 Mpc away. The largest known structures are a "chain" located in Aquarius, where 20 rich clusters form a 500 Mpc long "string" pointed radially away from us, and the "Sloan Great Wall," discovered in the SDSS redshift survey and also present in the 2dF data (see Fig. 2.3). It lies behind the "Great Wall" at a distance of 300 Mpc, extending nearly 450 Mpc (roughly from the head of Hydra to the feet of Virgo).

The list of Bahcall and Soneira [61] contains 18 superclusters, identified by applying a spatial correlation function on a sample of 104 Abell clusters in the distance class $D \leq 4$. A recent catalog by Zucca et al. [62] lists no less than 76 superclusters. They surround sparsely populated regions (voids) of comparable sizes creating the cellular-like morphology of the universe on large scales. Dark matter appears to be associated with all kinds of structures. There is a direct correlation – as the greatest amounts are associated with the largest scales (Fig. 2.15).

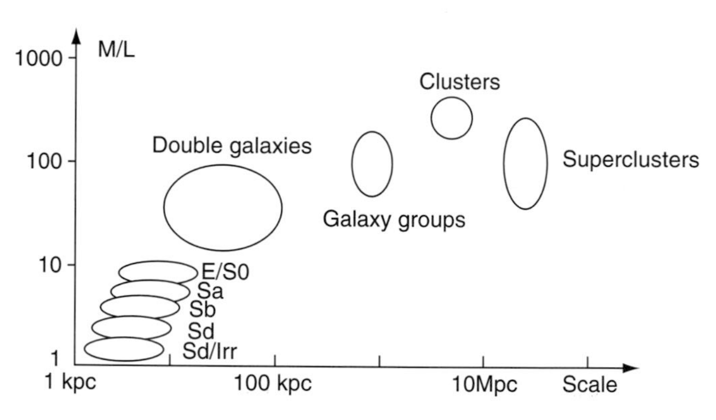

Fig. 2.15. Mass-to-light ratio for galaxies, pairs, groups, and clusters of galaxies

Close to the Edge

The Hubble Deep Field shows in addition to a few faint foreground stars almost 3,000 galaxies! An even deeper image was made in 2003 using HSTs *Advanced Camera for Surveys* (ACS), imaging an 3′ × 2′ area 1° southeast of M 31 with a limiting magnitude of 30.7 mag (exposure time 84 hours). Thousands of remote galaxies are shining through the faint halo stars of the Andromeda Nebula (Fig. 2.16). If one extrapolates the number of galaxies visible in the "deep fields" to the whole sky, there must be on the order of a trillion galaxies (each containing hundred billions of stars)! It staggers the mind just how immense the visible universe is.

Fig. 2.16. The Hubble Deep Field near M 31

Chapter 3

Catalogs, Data, and Nomenclature

Messier, Herschel, Caldwell, NGC/IC

The Messier catalog is the standard reference for any deep sky berserker [63–66]. In its extended version it contains 110 objects and is a quite inhomogeneous sample of non-stellar objects: galaxies, open and globular clusters, planetary and galactic nebulae. "M" is still the primary designation for bright deep sky objects, both in the amateur and professional scene. The Messier catalog is not sorted according to right ascension. The number of galaxies is 40, of which 16 belong to the Virgo Cluster. A few of the identifications remain controversial. In the case of M 102, some authors state an identity with M 101, a bright face-on spiral in Ursa Major, while others identify M 102 with the edge-on galaxy NGC 5866 in Draco. We here follow the latter view, in controversy to the recent statement that the mystery is "solved" [232]. Subject to some discussion is the identification of M 91 with NGC 4548, a relatively faint galaxy in Coma Berenices. Most Messier objects only need a binocular for detection.

The Messier catalog ignores many objects of comparable brightness. The Caldwell catalog, created by the English Amateur Patrick Moore [67,68], tries to fill the gap, listing the 109 best non-Messier objects. The C-number is introduced, referring to the full name Caldwell-Moore. The catalog is sorted by declination, starting with the North Pole region and contains 25 galaxies. Examples are the C 65, the "Silver Dollar Galaxy" NGC 253 (Fig. 3.1) and the faint companions NGC 147 (C 17) and NGC 185 (C 18) of the Andromeda Nebula, both located in Cassiopeia. The Caldwell catalog has not reached the popularity of Messier's. All Messier and Caldwell objects are visible in telescopes of 6–8 in. aperture.

The next step in the catalog hierarchy is marked by the *New General Catalogue* (NGC), which is the standard reference for deep sky objects of moderate brightness [69]. For all startled by a leap to 7,840 objects, there is an intermediate step: the "Herschel 400" list, published by the *Astronomical League* [70]. This list presents a selection of the best objects found by William Herschel, which includes 325 galaxies. This particular list was designed to be observed with a 6-in. scope under dark skies, though an aperture of 10–12 in. is recommended. The complete Herschel catalog contains 2,515 objects, of these 2,073 are galaxies.

The NGC was first published in 1888. Its appendix, the *Index Catalogue* (IC), adding another 5,386 objects, was published in two parts (IC I: 1895, IC II: 1908). Concerning format and data, the original NGC/IC is pretty out-of-date. The coordinates refer to the equinoxes 1860.0 and 1900.0, respectively. Instead of the declination the catalogs give the

Fig. 3.1. The brightest non-Messier galaxy: NGC 253, a member of the Sculptor group

obsolete "north pole distance" (NPD). The description of the objects (brightness, size, shape, nearby stars) is coded. The 13,226 NGC/IC objects present a mix of all classes of nonstellar objects. Due to the many observers and different instruments contributing to the catalog, the compilation is very inhomogeneous. This is not true for the subset of the objects found by the Herschels, using mainly an 18.7 in. reflector. The NGC/IC contains around 10,000 galaxies. Most of the NGC galaxies lie in the magnitude range 13–15 mag and should be visible with 14–16 in. telescopes. The IC galaxies were mainly discovered by photography (especially those of the *Second Index Catalogue* with numbers above 1,530) and are typically 1–2 mag fainter; in some cases 18 in. or more aperture may be needed. One of the faintest objects is the dwarf galaxy IC 4107 with $m_{pg} = 18.5$ mag (Fig. 3.2).

The original NGC/IC contains numerous errors. A considerable fraction of all of the entries does not correspond to any real objects [71]. There were various published corrections, but no general revision until 1973. The *Revised New General Catalogue* (RNGC) by Sulentic and Tifft [72] was the first attempt to clean the data, using the POSS. Due to an enormous time pressure; the resulting publication was in some ways *worse* than the original. Many known corrections were ignored – and some new errors created! At places, where no object could be found (due to bad historical data), an RNGC-number was assigned to the nearest "anonymous" object, visible on the POSS – even some plate faults are now carrying an RNGC-number! The RNGC also used an unusual equinox: 1975, which is a useless compromise between 1950 and 2000. Instead of morphological data, RNGC objects are divided into seven classes; magnitudes are adopted from RC2, CGCG, or MCG (see below). Each object carries an NGC-like coded description, derived from the visual appearance on the POSS.

Fig. 3.2. IC 4107 in Coma Berenices, one of the faintest NGC/IC objects

In 1988 Roger Sinnott was the first to present an updated version of the entire NGC/IC. His *NGC 2000.0* [73] was published just in time for the first centennial of the NGC – perhaps another case of time pressure. Known corrections were ignored in favor of "modern" data (CGCG, MCG, etc.), which are not error-free. In spite of this, the work was more successful than the RNGC. Sinnott has precessed the original coordinates to the modern standard J2000.0 and sorted the results by right ascension. As a result of precessional shifts this version has disturbed the original order (by NGC- or IC-number), making it difficult to find a certain object. The data presented includes object class, constellation and a (slightly enhanced) coded description. Unfortunately, the columns "magnitude" (mostly photographic) and "size" (only the larger diameter is given) show many gaps.

The best modern source for NGC/IC objects is the *Revised New General Catalogue and Index Catalogue* [74]. This catalog is based on the historic data, taking into account all published corrections. Many "puzzles" still remain to be resolved. In critical cases visual observations have been made, to simulate the historic discovery conditions. The credit goes to the international NGC/IC project, a team of both amateur and professional astronomers [75]. The revised catalog presents the latest data for each object, including proved identifications and cross references with a large number of modern object-specific catalogs. Beside the classic NGC/IC objects, additional objects as designated in the literature by a suffix A, B,... (e.g., NGC 1023A), or components of multiple galaxies are included. This leads to a total number of 14,000 entries. All coordinates were remeasured, using the *Digitised Sky Survey* (DSS), with a precision of 1–2″. The data (for existing objects) have been fully updated and include the constellation, magnitude (*B*, *V*, *V′*), diameters (*a*, *b*), position angle and Hubble classification type.

Catalogs of Galaxies, Groups, and Clusters of Galaxies

All catalogs described above contain a mix of different classes of nonstellar deep sky objects. In comparison object-specific catalogs refer to a special class, e.g., galaxies. We must distinguish between "general catalogs of galaxies" and those featuring certain selections: type, area, object parameter, spectral band, or hierarchical structure (pairs, groups, clusters). It is easy to see how such diversity produces a confusing nomenclature. A large collection of catalogs can be found at the *Centre des Donneés Astronomique de Strasbourg* (CDS; see Appendix).

What is a perfect galaxy catalog? It must be founded on precise definitions (selection criteria), which the objects must meet to a certain limit of accuracy. All depends on the quality of the basic data. Essential guidelines are the strictness of the definitions, completeness, homogeneity, and error-freedom. Only a few catalogs are compatible with such high standards. It is often the case that a listed "object" does not meet the definitions to be included, or is identical to another one, or even does not exist. Thus the number of "real" objects is a mystery in many catalogs. Thus, facing the doubtful nature of many catalogs, it is advisable to use the term "entries" instead of "objects." For scientific use such high standards are necessary. For amateurs it is in most cases sufficient to know that the database is more or less "reliable."

General Catalogs of Galaxies, Surveys, and Databases

The classic general catalogs of galaxies are the *Catalogue of Galaxies and of Clusters of Galaxies* (CGCG, 1961–68) by Zwicky and the *Morphological Catalogue of Galaxies* (MCG, 1962–74) by Vorontsov-Velyaminov. Both catalogs are based on visual inspections of the POSS, and thus are not "general" in the pure sense; the CGCG features the northern sky, the MCG covers parts of the southern hemisphere too.

Let's begin with a few remarks on the POSS and its southern complement. The first survey (POSS I) was made between 1950 and 1958 with the 48 in. Palomar-Schmidt ("Oschin Telescope"). It comes in two versions, using red- and blue-sensitive plates ("E" and "O"). In 1970, when better emulsions were available, a new two-color survey was started with the same telescope covering the sky down to –33° declination. The resulting POSS II shows a better resolution and a fainter magnitude limit (around 22 mag) [76]. The southern hemisphere was covered with the ESO Schmidt telescopes. All plates were later scanned for the *Digitised Sky Survey* (DSS). There are four versions: DSS I red/blue and DSS II red/blue, depending on the various plates used.

The CGCG lists 29,378 galaxies and 9,133 galaxy clusters north of –3.5° declination. The only data given are coordinate, (photographic) magnitude, and a not always reliable cross-reference to the NGC/IC. The magnitude limit is m_{pg} = 15.7 mag. The CGCG is considered to be essentially complete to around 15 mag. A typical designation reads CGCG 335-17 (which is M 31). The first number indicates the POSS field (plate number), running from 1 to 559; the polar field (370) is divided into parts A and B. The second number (separated by -) counts the galaxies in the field. The position accuracy is 1′. There is a revised version: the *Updated Zwicky Catalog* (UCZ) by Falco (1999), containing the positions of approximately 19,000 CGCG galaxies with an accuracy of 1–2″.

The MCG lists 31,917 galaxies north of –45° declination. The data about form and size are detailed, but difficult to decode. The magnitudes result from rough estimates on the blue POSS (O-magnitude) are only given to identify the galaxy. Errors up to 2 mag are possible. The galaxy diameters are much better; sizes derived from the red and blue POSS images are given. Vorontsov-Velyaminov states that their catalog is complete to 15 mag. This magnitude limit is rather indistinct as there are galaxies listed down to 20 mag. The designation for M 31 is MCG +7-2-16; the first number is a declination zone (–6 to +15), the second number is a field number in this zone, and the third counts the galaxies in the field. Beside the first sign (+,–), the two following (–) are for separation only. The position accuracy is only 1–2′. In 1998 Corwin published *Accurate Positions for MCG Galaxies*, listing 4,741 galaxies.

Unlike the CGCG and MCG, the *Uppsala General Catalogue* (UGC), published by Peter Nilson in 1973, is a catalog with a rather strong definition [77–79], and is based on the POSS too. It includes all galaxies north of –2.5°, larger than 1′ or brighter than m_{pg} = 14.5 mag (regardless of their size). The UGC lists 12,940 objects; an appendix contains 19 additional objects, marked by "A" (e.g., UGC 5854A = NGC 3357). The catalog gives type, magnitude (CGCG), size (from blue and red POSS) and position angle. In 1974 a supplement (UGCA) was published [80], listing 444 selected objects of special interest, south of the declination limit (e.g., UGCA 366 = M 83, Fig. 3.3).

Fig. 3.3. M 83 (UGCA 366) in Hydra

The *Catalog of Principal Galaxies* (PGC), published by Paturel et al., is a compendium (database) of a large number of galaxy catalogs (e.g., MCG, CGCG, UGC). The first version (1989) lists 73,197 objects, the second (1996) 108,792. The project was continued as LEDA (*Lyon Extragalactic Database*). It now contains 3 million entries. Unfortunately, both designations, LEDA and PGC, are in use. As most data are adopted from the source catalogs, their quality depends on the particular origin. But the LEDA team tries to enhance the data quality, which is a time-consuming task. The PGC gives type, magnitude, size, position angle, and cross identifications. If available, other data are listed, e.g., radial velocities. The position accuracy is (in the unrevised parts) 1–2′.

The *Third Reference Catalogue of Bright Galaxies* (RC3) by de Vaucouleurs (1991) contains data about 23,022 galaxies (selected from the PGC). Concerning the quality, types and completeness of the data, the RC3 is unexcelled. It is the prototype of a well-defined, homogenous magnitude/diameter system, presenting B_T, $(B–V)_T$, B_{25}' and the standard diameters. It includes error estimates for most quantities. The RC3 is the standard reference for the de Vaucouleurs classification. The relevant literature (thus "reference catalog") for each galaxy is compiled. It is much larger and more complete than its predecessor. The RC2 (1976) contains "anonymous" objects (A), which stands for "non-NGC/IC." A few are already present in the RC1 (1964), e.g., A0058 (RC1) = A0057-33 (RC2), the Sculptor system.

Much smaller, but equally valuable, is the *Revised Shapley–Ames Catalog of Bright Galaxies*, published by Sandage and Tammann in 1981 [81]. In addition to the 1,246 galaxies of the original catalog, it contains a lot of fainter objects, making it complete to a magnitude limit of B_T = 13.2 mag. The position accuracy is 0.1′.

A typical example of a survey ("Durchmusterung") is the *ESO/Uppsala Survey of the ESO (B)-Atlas* published in 1982 [82]. Based on the ESO-Schmidt plates, all nonstellar objects larger than 1′ and south of –17.5° are listed. Most of the 18,438 entries of the catalog are galaxies; a typical designation is ESO 29-G21 (Small Magellanic Cloud). The first number denotes the field (plate). G denotes the object class "galaxy," IG means "interacting galaxy" (these letters are omitted here in most cases). Other classes are PN – planetary nebula, SC = star cluster (some asteroids are listed too). The catalog gives no magnitudes. Exact photometric data on 15,467 galaxies can be found in *The Surface Photometry Catalogue of ESO/Uppsala Galaxies* (ESO-LV) [83]; where "LV" stands for the authors Lauberts and Vilcnk. It gives B_T values with an error of 0.1 mag.

Another valuable source for southern galaxies is the *Southern Galaxy Catalogue* (SGC) by Corwin and de Vaucouleurs (1985) [84]. It contains 5,481 objects south of –17° declination and larger than 1.5′. Included are types according to the de Vaucouleur classification, angular diameter, but no magnitude. Positions are accurate to around 0.1′. The gap between the CGCG- and ESO catalogs (reaching from –3.5° to –17.5° declination) is filled by the *South-Equatorial Galaxy Catalogue* (ESGC) of Corwin and Skiff (2000), listing 3,304 Galaxies between 3° and –20° with coordinates (precision 2″), type and size [85].

Modern surveys use the method of "automatic plate measuring" (APM), where raster data (scanned plates or DSS images) are converted into object data. The software is able to filter out certain types of objects (stars, galaxies). One of the first projects was the Hubble *Guide Star Catalog* (GSC). Besides the 15 million stars it also contains 3.4 million "non-stars," which are mostly galaxies. The GSC must be treated with care. The process can create "stars," where originally was an asteroid, a galaxy or even a plate flaw. Although special software has been developed to trying to detect and eliminate such errors, new trouble can sometimes arise by deleting real stars. Examples of galaxy catalogs that are based on automatic plate measuring include Lick Observatory's "North Proper Motion Program, 1st List of Galaxies" (NPM1G; 50,517 entries), the *APM Bright Galaxy Catalog*

Fig. 3.4. Distribution of galaxies on the sphere from the Automatic Plate Measuring (APM) catalog

(APM; 14,681 galaxies; Fig. 3.4), and the *Galaxy Properties at the North Galactic Pole* (NGP9; 36,402). The largest survey currently in progress is the *Sloan Digital Sky Survey.* Note that magnitudes and sizes derived automatically must be used with caution. For example, it is known that the NPM1G derived magnitudes can be off by 1 mag. As cross-references has not been given in these surveys and there is often the problem of identification with objects from the standard catalogs (which show a much lower position accuracy).

Cataloged objects coming from a variety of different sources can be arranged in a "database." The Messier- and NGC/IC catalogs are early examples. The first large printed database was Dixon's monumental "Master List" [86]. For amateur galaxy observations there are two recommendable printed databases: the *Sky Catalogue 2000.0, Vol. 2,* and the *Deep Sky Field Guide* (DSFG). The former catalog presents a good compilation of modern data, listing a, b, B_T, and for some galaxies also V_T. It includes also quasars. The DSFG (maybe the last printed database) lists all objects with detailed data (a, b, V_T, V', type), plotted in the *Uranometria 2000.0.* The trend goes to digital databases, available on CD-ROM or in the Internet. Examples of the latter are the *NASA Extragalactic Database* (NED), SIMBAD (specialized on galactic objects), or LEDA [87]. An interesting database compiled by amateurs, comes from the American *Saguaro Astronomy Club* (SAC). A large collection of databases can be found at the *Data and Archive Center* of the *Hubble Space Telescope Science Center* [88].

All modern sky mapping software (e.g., Guide, The Sky, Megastar) uses digital databases. Such programs might terminate the era of large-sized printed catalogs and atlases. But be aware that "modern" or "digital" does not automatically imply "correct"!

Catalogs of Special Types of Galaxies

The following are a few examples of catalogs that feature special types of galaxies (catalogs presented here can be found at the CDS; see Appendix). Let's start with dwarf galaxies. A classic source is the *David Dunlop Observatory* catalog (DDO; 243) by Sidney van den Bergh [89]. Slightly larger is the *Catalog of Dwarf Galaxies* by Karachentseva (KDWG; 260). Much less known are the collection of spherical dwarf galaxies presented by Mailyan (104). The small but celebrated Holmberg objects (I–IX) are dwarfs located outside the Local Group. Specialized on nearby galaxies, which are not only dwarfs, is Tully's *Nearby Galaxy Catalog* (NBG; 2,367) [90].

The primary source for fans of edge-on or superthin galaxies is the *Revised Flat Galaxy Catalog* (RFGC; 4,444) by Karachentsev (superseding the FGC). A great variety of peculiar features and systems are presented in Arp's *Atlas of Peculiar Galaxies* (Arp; 338), the *Atlas of Interacting Galaxies* (VV; 852; in two parts) by Vorontsov-Velayminov, and in Zwicky's *Catalogue of Selected Compact Galaxies and of Post-Eruptive Galaxies* (1 Zw to 8 Zw; approx. 3,000 objects). The southern sky is represented in the *Atlas of Southern Peculiar Galaxies and Associations* by Arp and Madore (AM; 6,445 entries) [91]. The special case of polar ring galaxies (Fig. 3.5) is presented in the *Atlas of Polar Ring Galaxies* (PRC; 157).

Fig. 3.5. The polar ring galaxy NGC 4650A in Centaurus

The standard reference for AGN, quasars and BL Lacertae objects is the list of Veron-Cetty and Veron, which is updated roughly once per year. The present version (11th edition, 2003) lists 15,069 AGN (among them 11,777 Seyfert galaxies), 48,921 quasars and 876 BL Lacertae objects [92]. The magnitudes are of type B, V, R, or photographic (POSS). Not up-to-date but still useful by its large amount of information presented, is the *Revised and Updated Catalog of Quasi-stellar Objects* by Hewitt and Burbidge (1993) with 7,315 objects [93].

Many quasars and AGN are cataloged as "faint blue stars." The relevant lists are: Humason and Zwicky (HZ, 48), Tonanzintla (Ton; 419), Usher (US; 2,363), Palomar-Haro-Luyten (PHL; 8,725), Palomar-Berger (PB; 9,495) or Luyten Blue Star (LB; 11,444). Searching for emission-line objects or galaxies with strong UV radiation also led to many new objects, listed in: Markarian (Mrk; 1,515; also known as *First Biurakan Survey*, FBS), Arakelian (Akn; 591), Kazarian (Kaz; 466), Haro (44), Palomar-Green (PG; 1874). Less known are: Wasilewski (WAS, 96), *Second Biurakan Survey* (SBS; around 1,700), University of Michigan (UM; 655) or the *Kiso UV-Galaxy Survey* (KUG; 8,968). Often quasars are simply desgignated by "Q" plus rough coordinates. By the way: it is still a miracle, what is meant by "Q 6188" in the Uranometria I (chart 261 and 262). This is the galaxy Mrk 960 (PGC 2845), but the "quasar-like" designation remains unexplained – perhaps an error.

Catalogs of Pairs, Groups, and Clusters

The above-mentioned catalogs of Arp, Vorontsov-Velyaminov, and Zwicky contain a large number of pairs and groups of galaxies that show signs of interaction. More obscure sources for galaxy pairs and triples are the *Catalog of Isolated Pairs of Galaxies* (KCPG; 603) and the *Catalog of Isolated Triplets of Galaxies* (KCTG; 84), both published by the Karachentsevs, following their *Catalog of Isolated Galaxies* (KARA; 1106 galaxies). A bit older, but still valuable are the lists of Klemola, Page, Rose, Snow, Turner, Eichendorf and Reinhardt, and the classic publication by Holmberg, "A Study of Double and Multiple Galaxies" (1937), listing 827 groups.

Often groups are not looking "isolated," the members are distributed over a large area on the sphere. This is the case for loose, nearby groups, which can only be detected by the similar redshifts of their members. A comprehensive source is the *Lyon Groups of Galaxies* (LGG; 485). The reverse of this is the occurrence and characteristics of "compact groups." The two most prominent sources are the highly popular *Atlas of Compact Groups of Galaxies* (HCG; 100) [94] by Paul Hickson, and the catalog of *Compact Groups of Compact Galaxies* (Shkh; 377) by Romela Shakhbazian and her co-workers.

The classic source for "poor clusters" and the intermediate aggregates between groups and clusters, are the Yerkes Observatory lists (MKW, AMW) [57], which has been combined and extended with the *Catalog of Nearby Poor Clusters of Galaxies* by White et al. (WBL; 732). The standard source for galaxy clusters is still Abell's catalog, titled *The Distribution of Rich Clusters of Galaxies* (A, ARC, or AGC; 2,712 entries; note that "Abell" is ambiguous, as there is an Abell catalog of planetary nebulae). In 1989, Abell, Corwin, and Olowin (ACO) listed another 1,174 clusters that are found in the southern sky. The 9,133 clusters found by Zwicky are published in the *Catalogue of Galaxies and Clusters of Galaxies* (CGCG).

There are special catalogs of individual galaxies in clusters too. An early example is Alan Dressler's *Catalog of Morphological Types in 55 Rich Clusters of Galaxies*, listing around 6,000 galaxies. For the most prominent clusters there are works by Rood and

Baum or Doi et al. for the Coma Cluster, the *Virgo Cluster Catalog* (VCC) by Bingelli et al., the *Fornax Cluster Catalog* by Ferguson, or Dickens' catalog for the Centaurus Cluster.

Cross Identification, Names

Often an object appears in different catalogs, being observed or discovered in various programs or surveys. The main problem is to recognize objects that bearing different designations as being identical. This process is called "identification." Identical objects can be marked by a "cross-reference." This looks easy, but can be the cause of much confusion. If there are reliable data, the problem is minimal, but can become difficult, even unresolvable in the case of catalog errors. There are numerous instances in which the "true" identity was not realized for a long time.

For a bright galaxy, cross-references will produce a long list of different designations. This leads to the question, which is the "primary name"? Fortunately, there is a "canonical" priority sequence: M, NGC/IC, UGC, MCG, CGCG, PGC, and so on. This will be explained by two examples.

NGC 4517 is a bright spiral galaxy (V = 10.4 mag) in Virgo (Fig. 3.6), and was not noticed by Messier. The cross-reference sequence reads in this case as: NGC 4437, UGC

Fig. 3.6. NGC 4517 in Virgo, with its (optical) companion NGC 4517A

7694, MCG 0-32-20, CGCG 14-63, PGC 41613, FGC 1455, KCPG 344B, IRAS 12301+0023, UM 505. The second NGC-number comes from an internal identity, due to a double discovery that was not recognized in the original historic catalog. The further designations are typical for a bright northern galaxy. FGC refers to a "flat galaxy" (the objects have dimensions of 10.2′ × 1.7′). The KCPG-number shows the object to be a member of an "isolated pair of galaxies" (the other component is KCPG 344A = NGC 4517A). The IRAS- and UM-designations indicate that the galaxy is both an infrared source and an emission-line object. As there is no entry in the *Virgo Cluster Catalog* (VCC), NGC 4517 does not belong to this cluster. To sum up: just looking at the different designations already yields a lot of information about the object.

The second example is NGC 5421, a spiral galaxy ($V = 13.4$ mag) in Canes Venatici (Fig. 3.7), identical with UGC 8941, MCG 6-31-45, CGCG 191-33, PGC 49950, IRAS 13594+3404, KCPG 407B, Arp 111, VV 120, Mrk 665, and I Zw 78. Up to the KCPG designation all looks similar. The Arp- and VV-numbers point to a certain "peculiarity," probably due to an interaction with the companion (KCPG 407A). The Mrk- and Zw-designations indicate that the object is a galaxy with UV-continuum and "post-eruptive" (Zwicky's criteria are not very precise).

For quasars the naming problem becomes truly difficult, as the primary designation is not obvious. Thus, many quasars live with different names making it difficult to recognize an identity. A classic example is Ton 599, also known as 4C 29.45. Fortunately, comparing coordinates and redshifts (which are usually accurate) usually helps to clear up the situation. In some cases, there is a "historical" name: there is no doubt about 3C 273,

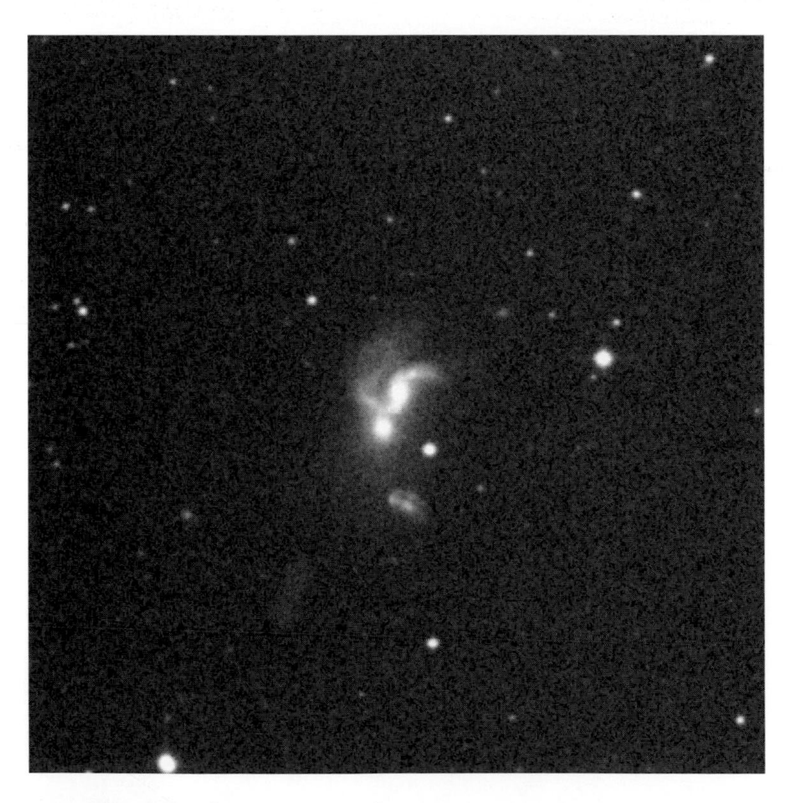

Fig. 3.7. The interacting galaxy NGC 5421 in Canes Venatici

which is #273 in the third catalog of radio sources detected at Cambridge. The other names that cannot confuse include: H 1226+023, 4C 02.32, PKS 1226+02, ON+044, NRAO 400, DA 324, MSH 12+08, PG 1226+023, PGC 41121 (and 42 more!). Except PGC (galaxy) and PG (bright quasar), all these are radio source designations.

Many prominent, remarkable, or historically interesting galaxies bear proper names. Due to the popularity of the Internet there has been an inflation of new names, but fortunately there is a common treasure of "classic" ones. Here are a few of the best known examples: "Sombrero Galaxy" (M 104), "Integral Sign Galaxy" (UGC 3697), "Black Eye Galaxy" (M 64), the "Siamese Twins" (NGC 4567/68), "Coddington Nebula" (IC 2574), and the "Whale Galaxy" NGC 4631 (Fig. 3.8).

Data Quality

When dealing with catalogs, the term "error" can be interpreted in many ways. The main categories are: nonexisting objects, wrong identifications, incorrect object types, data errors, or even simple typos. Often one is confronted with a complex combination, which is difficult to resolve. Errors have been reproduced from one catalog and then transferred to another, and often the classic sources (M, NGC/IC) are involved. When a problem

Fig. 3.8. The "Whale Galaxy" NGC 4631 in Canes Venatici and its companion NGC 4627

arises, try to consult a variety of different sources. Let's take look on a few examples for each category, starting with the simplest – the typo.

Sometimes a typo is obvious, such as a wrong declination sign. Often it is not, as in case of digit or letter errors (CGCG 387–18 = NGC 7628, should read NGC 728). For most catalogs lists of errors were published. Ignoring them can cause much trouble, as in the case of "Copeland's Septet" in the RNGC (Fig. 3.9) [95]. This prominent compact galaxy group was flagged "nonexistent" there. The authors had precessed the original NGC-coordinates – which were known to be wrong! This leads to a blank field.

Some objects are really missing. Perhaps they exist at a different, but unknown position. Examples are the "galaxies" MCG 8-9-2, UGC 12154, or ESO 139-57. True nonexistent objects are the photographic "plate flaws" UGC 5196, PGC 6622, or PGC 24675.

As already mentioned, there are many examples, where galaxies (e.g., when looking peculiar) are confused with other classes of deep sky objects and vice versa (Table 3.1). Classic cases are variable compact galaxies (quasars, AGN) that were cataloged as "variable stars" [96]. Sometimes parts of galaxies (e.g., bright outer HII regions) were recognized as independent objects, as in the case of NGC 5906 (see *NGC 2000.0*), which is a bright knot in the large edge-on-galaxy NGC 5907.

Wrong or incomplete data can cause further identification problems. Approximately 5% of all NGC/IC objects are identical, e.g., NGC 3384 = NGC 3371 (galaxy in Leo), NGC

Fig. 3.9. "Copeland's Septet" in Leo, a case of trouble in the RNGC

Table 3.1. Examples of incorrect classes of deep sky objects

Object	Incorrect class	Correct class
UGCA 415	Galaxy	Planetary nebula (Abell 65)
NGC 2242	Galaxy (CGCG 204-5)	Planetary nebula (PK 170+15.1)
IC 3568	Galaxy (UGC 7731)	Planetary nebula (PK 123.6+34.5)
IC 4677	Galaxy (MCG 11-22-17)	Knot in planetary nebula (NGC 6543)
Abell 76	Planetary nebula	Ring galaxy (IRAS F21274-0301)
PK 248+8.1	Planetary nebula	Galaxy (ESO 495-21)
PK 137+16.1	Planetary nebula	Nearby galaxy (Cam A)
NGC 2296	Galaxy (MCG -3-18-3)	Molecular cloud (IRAS 06464-1650)
UGC 11668	Galaxy	Galactic nebula (GN 21.00.3)
GN 6.32.9	Galactic nebula	Galaxy (ESO 557-6)
Sh2-191, Sh2-197	Galactic nebulae	Nearby galaxies (Maffei I, Maffei II)
IV Zw 30	Galaxy (Mrk 959)	Globuar cluster in M 31 (Mayall IV = G 219)
NGC 2537	Globuar cluster	Peculiar galaxy (Arp 6)

2947 = IC 547 = IC 2494 (galaxy in Hydra), or NGC 3497 = NGC 3525 = NGC 3528 = IC 2624 (galaxy in Crater). The reverse case is an incorrect identification. The *NGC 2000.0* lists NGC 4006 as "galaxy" and assigns it to be identical with IC 2983, which is listed as a star. The truth is: NGC 4006 is not identical with IC 2983 and the latter is not a star, but should be characterized as "not found."

General Literature, Sky Atlases, and Software

There is a large amount of both printed and digital material that are relevant for topics of "visual observing" and "galaxies." This includes books, magazines, printed sky atlases, software, and websites. We will only mention some important facts concerning such sources here. References can be found in the appendix ("General Literature," "Digital Sources").

Printed Sources

Books, containing information on observing deep sky objects are quite numerous [97–99]. Galaxies, perhaps the most popular object class, are described in these works but must share space with the other classes. The same is true in the astronomical magazines. Unfortunately, there is no commercial magazine for deep sky observing (there are a few, published by astronomy associations). Three popular publications were terminated, *The Observer's Guide* (1987–1992), the *Deep Sky Magazine* (1982–1992), and *The Deep Sky*, National Deep Sky Observers Society (NDSOS) in 2004. Fortunately, there are published extracts of the best articles [100]. It is worth to looking for old issues. Their content, especially when concerning galaxy observations is still useful. The objects are the same and the visual observing techniques have not changed much over the years. Rare magazines or out-of-print books may be found in the Internet (www.abebooks.com, www.ebay.com).

The classic chart for deep sky observing is the printed sky atlas. It should "fit," the aperture used in terms of number of deep sky objects and limiting magnitude. The magnitude limit should be comparable to the object's magnitude or not more than 2–3 mag brighter. For the Messier and brighter NGC galaxies, a sky atlas showing stars to around 8 mag is sufficient (*Sky Atlas 2000.0*). To find fainter NGC objects, stars down to 11 mag should be plotted (*Uranometria 2000.0*, Millennium *Star Atlas*). But in case of small, faint objects a sky atlas is in most cases not really sufficient for finding – nevertheless it does often show them! Additionally a more detailed, deeper finding chart is necessary. There are still more problems. If a right angle star diagonal is used, the charts cannot be inverted.

Sky Mapping Software, Internet Sources

Many problems often associated with sky atlases can be solved using a "sky mapping software" ("planetarium program"), which "visualize" the sky on the computer screen. Prominent examples are *Guide*, *Megastar*, *SkyMap Pro*, or *The Sky*. Are the classic sky atlases obsolete? If starhopping is the preferred method for finding objects, a printed sky atlas is still useful as an intermediate step for orientation. If a computer is not present at the observing site, then the detailed finding charts must be printed out in advance.

All modern sky mapping software uses large databases to generate their charts. The standard source for stars is the *Guide Star Catalog* (GSC), often enhanced by the Tycho- and Hipparcos catalogs. The GSC contains stars as faint as 15 mag. Even deeper is the USNO catalog, presenting a very deep limit that goes down to 20th magnitude [101]. Unfortunately, USNO magnitudes are not often highly reliable and can be in error by 1–2 mag. This massive amount of data can now be stored on larger hard drives, or you can have "fun" managing 11 CD-ROMs – and feeling much like a disk jockey.

As mentioned above, the deep sky databases installed in sky mapping software are not always reliable. Trouble arises, if the "digital" sky differs from the "real" sky. Fortunately, most programs offer the possibility of loading the *Digitised Sky Survey* (DSS) as a background image. This helps to clear the situation. The DSS can come via Internet or from CD-ROM (*RealSky*). The future has already begun: the *Sloan Digital Sky Survey* (SDSS) now offers 100 million objects down to 23 mag! At least 25% of the sky will be digitally mapped at this level of detail, resulting in 15 terabytes of data.

Searching for published papers with astrophysical content (e.g., concerning all aspects of galaxies) the best source is the *Astrophysics Data System* (ADS) at Harvard University. It offers a large number of abstracts and scanned articles of all major professional journals. The latest papers, submitted for publishing, can be found on a "preprint server" (see Appendix).

Section II

Technical Aspects on Observing Galaxies

In the last section, we talked about the physical nature of galaxies and clusters, including the relevant information on data sources and their quality. This section presents the technical aspects of observing, both instrumental and physiological [102]. Descriptions of telescope types, like the standard instrument for visual observations, the Dobsonian, are omitted. You can find enough informations in the published literature or the internet. But we will focus on necessary tools like finders, eyepieces, or filters here. Beside the many instrumental aspects, galaxy observing is also a matter of vision techniques. A bit of theory is necessary to discuss questions like: What are the essential optical quantities? How to use averted vision? What is the relevance of aperture and magnification?

OK, let's talk a little about "aperture-fever." Independent of the telescope size, you will always find objects looking similar to those observed in a binocular. Sure, it is a benefit of a large aperture to delve deeply into the structure of bright galaxies (see, e.g., the observations of Ron Buta [103]). But under the right conditions even a small telescope can discern considerable detail. It can be quite fascinating to discover a certain Messier or even NGC galaxy in a small instrument. And sometimes, in case of large, low surface brightness galaxies, a small aperture can be even better! There are challenging cases for every size and no serious visual observer would joke about small telescopes. It needs the same (or even more) degree of experience and observing technique to detect a 13-mag galaxy in a 4 in., than a 16-mag galaxy in a 20 in. In fact, there can be a "minimum-aperture-fever" too!

Chapter 4

Accessories and Optical Quantities

Eyepieces, Filters, and Optical Accessories

Eyepieces, Exit Pupil

A good eyepiece is essential for successful visual observing. The number of different optical designs and products is large [105]. Generally the quality requirements for eyepieces increase for telescopes with higher aperture ratios, because of their larger optical aberrations. More simple (cheaper) designs can be used for Newtonian or refracting telescopes with 1:10. Most eyepieces are built for systems that have fewer optical aberrations. A short-focus Newtonian or a Schmidt Cassegrain Telescope (SCT) often require well-corrected eyepieces for the best performance. The problem of the latter is spherical aberration, due to the curvature of the secondary mirror. To avoid frustration, try to test different eyepieces at your telescope before buying.

Is there an ideal eyepiece for observations of faint, small galaxies, large low surface brightness objects or clusters of galaxies? Obviously the qualities needed are: sharp image (even at high magnification), high contrast, and large apparent field of view. A single eyepiece, regardless of its quality, can impossibly manage this.

Let's start with focal length, magnification and exit pupil. There is a large range of focal length (f'); from 2.5 to 50 mm. Short focus eyepieces ($f' < 15$ mm) normally come with a diameter (barrel size) of 1¼ in. For larger focal length the 2 in. barrel size is available, which offers a larger field of view. Smaller telescopes are usually equipped with a 1¼ in. focuser. Reflectors of 10 in. or more should have a 2 in. focuser. To use a 1¼ in eyepiece in a 2 in. focuser, a reducer is necessary.

Which focal length is needed to realize typical magnifications between 50 and 500? This depends on the focal length of the telescope (f). Magnification is defined by $m = f/f'$. In case of an 8 in. SCT with $f = 2$ m, eyepieces between $f' = 4$ mm and $f' = 40$ mm are required. A 3.5 mm eyepiece at a 20 in. 1:5 Dobsonian would lead to a magnification of $m = 2{,}500$ mm/3.5 mm $= 714$, which is only useful under excellent seeing conditions.

Beside magnification the exit pupil is an important quantity for visual observing. At the focused eyepiece a bundle of parallel light rays exits towards the observer's eye, which is a reduced image of the telescope aperture. The bundle is visible as a bright round spot, which does not fill the whole lens (Fig. 4.1), best seen at daylight from a distance of around 30 cm (the spot moves if you move). Its diameter is called "exit pupil" p. To calculate p, the aperture number $N = F/D$ of the telescope is needed. The exit pupil (in mm) is then given by $p = f'/N$, with f' in mm.

Fig. 4.1. Exit pupil (white spot in the eyepiece)

An 8 in. SCT with $f = 2$ m gives $N = 2{,}000/200 = 10$. Using an eyepiece with $f' = 25$ mm the exit pupil is $p = 25$ mm$/10 = 2.5$ mm. Again using the same telescope, N is fixed, thus p depends only on the eyepiece focal length: p increases with larger f'. There is an alternative formula for p, using the magnification: $p = D/m$ (with D in mm). As D is a fixed value too, we see, that p decreases with larger m. This formula is most useful for binoculars, where $m \times D$ is given. Applied to our SCT, we have $m = 2{,}000/25 = 80$, getting $p = 200$ mm$/80 = 2.5$ mm (as above).

The apparent field of view ("fov," measured in degrees) defines the "space" seen in the eyepiece. A value of only 40°, typical for low cost eyepieces, creates a "tunnel view." This may be acceptable for observing the moon or planets, but not for galaxies. With a large value of 82°, observing is like a "spacewalk." The eye is unable to glance the whole field at once. An eyepiece with large apparent field forces the eye to re-focus when turning to a different part of the image. Though the apparent field may be up to 70 or 80°, the true field is dependant of the amount of magnification used. Also of note is that these impressive eyepieces can contain up to seven lenses. But every glass–air transition reduces the contrast, even with a good anti-reflection coating – so there will be a trade off for a wider field of view or *fov* vs. the contrast and sharpness of the image.

Some products offer a large eye relief (maximum distance between the eye and the eyepiece, while still seeing the entire field of view). This is an important tool for wearer of glasses [106]. To save the lens and to darken the view a rubber eye guard is helpful. In addition, blackened lens edges and anti-reflection threads deliver a maximum contrast.

Filters

The glow of the night sky caused by the unfortunate combination of light and air pollution is a major factor influencing the success of deep sky observing (Fig. 4.2). Note that the surface brightness of many faint, extended galaxies is comparable with that of a dark night sky. Thus light pollution will drastically reduce your observing program. Is larger aperture a solution? Unfortunately not: it even amplifies the bad situation in collecting both "good" and "bad" light.

Most of the sky glow comes from terrestrial (artificial) sources, mainly from street lights [107]. The two main types are the high-pressure sodium vapor lamp and the high-pressure mercury vapor lamp. The first (pink tint) emits in the range 550–630 nm, with several maxima. The second (blue/white tint) emits at 405 nm, 436 nm, and in the range 540–630 nm. A minor part of the sky glow comes from natural sources in the upper atmosphere, mainly airglow and auroral lines, due to the interaction of charged solar wind particles with air molecules of oxygen (555.7 and 630.0 nm) and sodium (589.3 nm).

A filter maybe helpful, but don't expect wonders when galaxies are your favorite targets. Note that a filter does not lighten the object, but darkens the background (which is actually a "foreground"). It is specified by its transmission curve and we generally distinguish between two types: broadband and narrowband. A broadband filter, e.g., light pollution reduction (LPR), shows low transmission in the wavelength region of the major light pollution but wide bandpasses of 100 nm or more otherwise. Narrowband filters, also called "nebular filters," e.g., ultra high contrast (UHC), are characterized by bandpasses of 30–50 nm around the most important emission lines: Hα (656.3 nm), Hβ (489.9 nm), and OIII (495.9 and 500.7 nm). It is ideal for enhancing the contrast in case of emission nebulae (HII regions) and planetary nebulae [108]. The extreme form is a "line filter," which shows a bandpass of only 10–15 nm centered on a single line (or narrow "double" in case of OIII).

Galaxies are, like stars, continuum sources, emitting at all wavelength. Thus a nebular filter makes no sense, with one exception: it can be useful for emission line objects in nearby galaxies. Extragalactic HII regions, like NGC 604 in M 33, are usually seen well with an Hβ-filter. Another example is IC 1308 in Barnard's Galaxy, where an OIII filter is useful (Fig. 4.3). Broadband filters like LPR or UHC will dim both the object and the background. But if the light pollution comes from a source with only a few lines or a narrow spectrum (e.g., a low pressure sodium vapor lamp), it can be blocked more effectively, however this requires the proper filter. For the common broadband filters the contrast enhancement in case of galaxies is only marginal. Anyway, in the case of low surface brightness galaxies like M 101 an LPR filter can be useful.

Fig. 4.2. Light pollution over a city

Fig. 4.3. The HII region IC 1308 = Hubble X in NGC 6822 (Ophiuchus)

These filters come in two standard 1 ¼ and 2 in.barrel sizes (2 in. filters are more than twice as expensive). The filter is screwed into the rear side of the eyepiece tube. It has a maximum effect only for rays entering perpendicular to the (plane) surface. At its position in the eyepiece, this is only valid for the central ray as all other rays are convergent. As the angular deviation depends on the focal ratio, the same filter works much better at a 1:15 refractor than at a Newtonian with 1:5. The spectral band is shifted up to 4 nm to the blue for high ratios, thus for a narrowband filter the transmission can be reduced significantly. A Barlow lens (see below), which decreases the ratio, may improve the situation [109].

Finding Tools

Two different finding tools will be discussed. If starhopping is practised, the standard is an ordinary finderscope. The most sophisticated method is "GoTo," using the aid of a sky-computer.

Naked-Eye Tools, Finderscopes

The simplest type of a finder lacks any optical components, thus no magnification, rotation or inversion is present. Most popular are "Telrad" and "Star-beam" [110]. The Telrad appears to project a red bull's-eye pattern (via a mirror) onto the night sky. Its brightness can be regulated to include fainter stars in the search. For the common star atlases there are Telrad stencils, fitting on the map. A digital variant is included in most star mapping software, plotting the Telrad field on the screen. A Star-beam can be used, if the telescope is too small for a Telrad. It appears to project a small red dot onto the sky, indicating the present orientation of the telescope. Without such devices positioning can be done via simple markers on the telescope tube.

Naked-eye finders are sufficient only to locate bright targets. But they are helpful as the first step in the finding procedure, followed by the optical finderscope. The position of the finderscope depends on the telescope, but should be near the eyepiece. This is the upper end of the tube for a Newtonian (close to the telrad) or the lower end for a refractor or SCT. In case of a large aperture SCT it may be comfortable to install two finderscopes, located in opposite positions. For a larger Dobsonians a second finderscope makes sense too, put at the lower end of the tube, near the rocker box. To locate a difficult field, the star chart (on a nearby table) must be consulted several times. No problem, if the elevation is low and no stair is needed. But at high elevation, you must climb the ladder for each trial. Thus the lower finderscope saves time and make things more secure.

It is important, that the finderscope is correctly aligned and focused before starting the observing session. This can be done already at dawn using a bright star. A good finderscope has cross-hairs illuminated by red light. If the brightness can be adjusted, be sure that a low value is used. A bright cross-hair affects the magnitude limit in the finderscope – and the dark adaptation. A standard finderscope gives an inverted image; with a zenit prism it is additionally mirror-reversed. This can cause much trouble for beginners, still learning orientation. To avoid confusion an Amici prism can be installed. Another way is using a “half” binocular as finderscope.

Unfortunately some commercial telescopes offer too small finderscopes. For starhopping 6 × 30 is not sufficient. Apertures of 40–60 mm are optimal. Also important considerations are magnification and true field of view. The magnification should give 2° or more and a reasonable exit pupil. A magnification of 10 is a good choice for a 50 mm lens. It is a benefit of large finderscopes that extended galaxies with low surface brightness (e.g., M 101; Fig. 4.4) can already be visible – sometimes even better that in the main tube!

GoTo

A finderscope or even setting circles may seem obsolete, especially if your telescope can do the job alone – by “GoTo.” It is a standard for modern computerized telescopes, e.g., altazimuth mounted SCTs. To start an observing session, the system must be set up with two reference stars. You then select an object from the database (or punch in its coordinates) and the telescope slews to the right position. All you have to do after that is to look through the eyepiece. Another way is offered by the use of standard sky mapping software. Many of these programs can control a telescope via a serial interface. Alternatively you can use the display as “digital setting circles.” This can also be realized for Dobsonians by installing encoders on both axes. With the aid of a computer the measured angles are transformed into equatorial coordinates. With the known object position you have to move the telescope until the display shows the desired values.

GoTo looks nice and easy, but there are some things to mention critically. If your telescope is not at a fixed place, you must carry a lot of equipment. The complete system is pretty expensive – you have to pay for comfort. What you can’t buy is experience: knowledge about the sky, its objects and physical relations. Thanks to the sky computer it is possible to find objects without knowing a single star or constellation. What do you learn about astronomy? Nothing. What is also lost is simply fun and the feeling of success – as a result of using your eyes and brain. Lack of experience can cause curious situations. Here

Fig. 4.4. A starhop away from Mizar: M 101 in Ursa Major

Fig. 4.5. The Draco dwarf spheroidal galaxy

is a short example. There are objects in the telescope database, which are senseless or extremely difficult targets. Try finding UGC 9749, better known as "Draco Dwarf" (Fig. 4.5). This is an extremely low surface brightness object that is among the most difficult targets in the sky, even for top visual observers – and for CCD imaging too! GoTo will move your telescope patiently to the exact field – but you will look in vain. A similar frustration arises, if your database contains nonexisting objects (which is not a rare case). Without proper experience, you may not find out why there is a blank field. Is this because the telescope is too small, the database incorrect, or perhaps the seeing too bad?

Try to learn something about the sky and its objects. Scan the skies freely and compare your view with the map. Spend time with starhopping and you will get more and more familiar with stars, constellations, and their objects. That's pure deep sky observing – leading to a strong, individual relationship between objects and observer. Your "personal" sky fills up with star trails, patters and mini-constellations. With time, you are able to notice even changes, due to moving objects, variability, or perhaps a nova!

GoTo can be useful – in the hand of an experienced observer. Observing programs, including many different targets, can be easily processed. The same is true for guided "star parties." But visitors are often more impressed, if their guide finds the objects "by hand." A final aspect: Beginners, joining the scene with a brand-new computerized SCT, expect to see colorful images of galaxies. Being not familiar with the (dim) reality, they get frustrated, which may rapidly terminate the new hobby. Anyone, who really wishes to learn the sky should follow the "natural instrumental sequence": starting with a binocular, followed by a small reflector to end with a larger Dobsonian. This guarantees an ongoing and fruitful connection with astronomy.

Chapter 5

Theory of Visual Observation

Let different people look at a galaxy with the same telescope under the same sky: they will see different things. One observer describes spiral structure and a lot of details; another detects only a faint patch of light. Why the large discrepancy? It's because observing is a matter of experience, vision technique and the right preparation. Generally the visibility of deep sky objects depends on factors of atmospheric, instrumental, technical, and physiological nature:

- Sky conditions (transparency, seeing)
- Aperture
- Clean, collimated optics
- Entrance/exit pupil
- Dark adaption
- Direct/averted vision
- Contrast
- Magnification

For galaxies, being faint objects with even fainter structures, all these factors are most relevant. So the situation differs in essential parts from observing the moon, planets or double stars.

Eye Sensitivity, Observing Techniques

The eye is the very treasure of the visual observer (Fig. 5.1) [236]. It is extremely sensible for brightness differences and motion, and can even accumulate light to a certain extend. The eye is an excellent contrast detector, especially at low light levels, being much better than film or even a digital camera. With experience and using the right techniques, one can push its performance to a very high level [111]. But there are also advancements in vision-correction surgery [237].

Cones and Rods, Dark Adaptation

The retina is equipped with two different kinds of light-sensitive cells: cones and rods. Cones are concentrated at the fovea, which is the center of vision at the optical axis of the

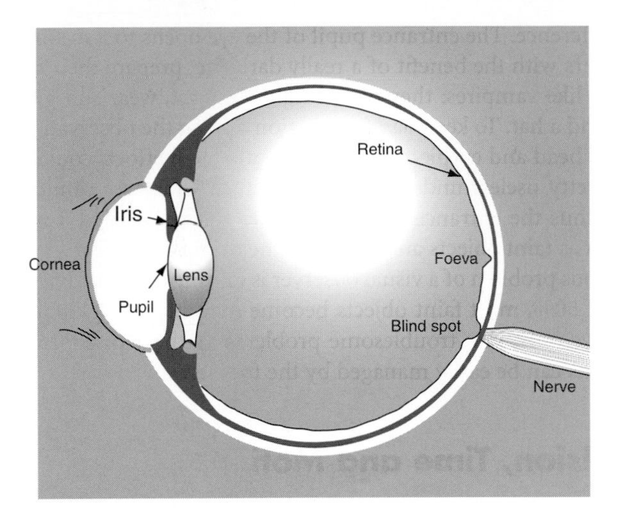

Fig. 5.1. The human detector of light: the eye

eye. They are high-resolution sensors that can detect color, but are less sensitive than rods. Rods are dominant off the optical axis. The areas of highest density are located at an angle of approximately 20° around the fovea. Rods can detect faint light, but only as gray intensities. Their resolution is less than that of the cones.

At daylight the cones are the primary detectors, seeing things sharp and in color (photopic view). This process is called "direct vision." Also at night bright objects are viewed with the fovea, e.g., planets with the naked eye or stars in the eyepiece. Some objects show even color in the telescope, like planets or high surface brightness planetary nebulae. With decreasing light, rods take over (scotopic view). Note that the wavelengths where cones and rods show their peak sensitivity are different [19]: 555 nm for cones (yellow light), 507 nm for rods (green light). Thus in the transition phase from cone to rod vision, the response shifts to shorter wavelengths (Purkinje effect). This is important for visual observing. Roughly speaking, for faint sources blue is better than red, for bright sources things are opposite.

On uses the term "visual" magnitude, because the peak transmission of the standard V-filter (550 nm) is pretty near the wavelength of maximum cone sensitivity. Thus V is nearly identical with "photopic magnitude." As most deep sky objects are faint, i.e., cones are not relevant, then the "scotopic magnitude," as defined by the maximum rod sensitivity, is the correct measure. For objects emitting in certain spectral lines, this value can be much different from visual magnitude. In comparison with V, an OIII source (planetary nebula) can be up to 1 mag brighter, a Hβ source (emission nebula) even 3 mag! The opposite is true for (normal) globular clusters, which are fairly "red" objects: the scotopic magnitude is 1 mag fainter than the visual. Galaxies, as continuum sources, are not much affected, but quasars can be: An object showing strong Hβ emission (486 nm) is intrinsically blue. At rest it would appear brighter than its V magnitude indicates. But in reality things are worse, due to redshift. With $z = 0.23$ the line would land at 600 nm, far beyond V. Thus the quasar appears fainter than in V. Unfortunately there is no general formula to calculate scotopic magnitudes from V magnitudes, given in the catalog.

Rod cells need a certain time to reach their maximum sensitivity. This process is called "dark adaptation." It's best to let the eyes adjust in the dark 30 min or more before starting an observing session. The time depends mainly on the condition of the observer and the light level difference. The entrance pupil of the eye opens to a maximum of 6–8 mm. Excellent observers with the benefit of a really dark site, prepare their session a few days earlier, behaving like vampires: they omit bright sunlight, wear sun-glasses with strong UV protection and a hat. To keep dark adaptation during the observation, it may be useful to cover both head and eyepiece by a black cloth. Such efforts could gain 1 or 2 mag. But all this is pretty useless under urban skies, since your eyes cannot reach full dark adaptation and thus the entrance pupils are never fully expanded. But what is lost? In reality, not much as faint objects are hidden in the bright night sky.

The most serious problem of a visual observer is loosing eye sensitivity. If the eye works only at a level of 50%, most faint objects become invisible. But perhaps the problem is limited to one eye only. Less troublesome problems are near or shortsightedness. Such optical aberrations can be easily managed by the focuser.

Averted Vision, Time and Motion

Very faint objects cannot be seen with the cones at the fovea, as they are not sensitive enough. Are they lost to the brain, if direct vision is impossible? This is certainly not the case. Faint light is a matter of rods. But how can one bring the targets into position? The answer is simple: try directing your vision between 6° and 16° away from the object by shifting your eye to an imaginary point. This technique is called "averted vision," guiding the light to the most sensitive parts of the retina. Averted vision, using the full rod power, will experience a gain of some four magnitudes or more over direct vision. But the price is lower resolution. Learn to look this way for faint targets – it is unfamiliar, but effective.

In principle any direction of averted vision is possible – except one. If you are a right-eyed observer and shift your eye to the left (towards the nose) the object's rays will hit the blind spot, where the optic nerve connects to the eye. Nothing will be seen, no matter how bright the source is. Left-eyed observers looking to the right will get the same negative result. Just the opposite is best. Whichever eye is used, avert your gaze in the same direction, always away from the nose: right eye to the right, left eye to the left.

What about the other directions: upwards or downwards? The area of the retina above the center of vision is a bit less sensitive. The same is true for shifting downward, but the sensitive area is smaller. So upwards is better – and this is the optimal way to practice averted vision in a binocular! If you avert your gaze to the right (or left), both eyes will shift – if not, you have a significant problem. Thus, one eye is positioned well, but the other will expose its blind spot.

In most cases the field of view in the eyepiece presents a mix of sources: brighter, fainter, point like, extended. If we directly glance at an (sufficiently bright) object, the cones are used through direct vision. But simultaneously the rest of the apparent field of view in the eyepiece is recognized too – naturally by averted vision. If not, we would only experience an extreme "tunnel view." Note, that this can happen if the apparent field of view of the eyepiece is too small (say 30°). With modern eyepieces, offering 60°, cones and rods are equally busy. At higher values (over 80°) the eye cannot catch the scene at once.

The eye–brain system can do other interesting things: aggregation and accumulation. In low light situations the eye is able to act as an "image intensifier." It uses an aggregation of neural transmitters, the "ganglion cells," which are able to combine adjacent rods. This feature is one of the reasons, why rods are more sensitive than cones – simply quantity against quality. The number of rods involved, depends on the area covered by the object on the retina. Its size is thus a matter of magnification (to be discussed in more detail later).

What is even less obvious is that the eye is able to accumulate light similar to film, but much less effective [112]. Thus taking ones time while observing an object is important. During the daytime, or looking at bright objects, the integration time is short. It is around 1/30 second, which is comparable to the time resolution of viewing. In case of faint light the eye–brain system can integrate the signal considerably longer: up to 6 seconds! Try scanning the object slowly, while holding the averted view for some 10 seconds and then turn to another part. Such intense viewing is quite tiring!

What is the limiting magnitude, under ideal conditions (transparency, seeing, dark adaptation, etc.) with a certain aperture using averted vision [113]? All depends on what is meant by the word to "see." The degree (or probability) of detection can be quantified by the fraction of time the object is really perceived compared to the total observing time. A value of 50% means, that it can be seen (with random interruptions) over 30 seconds in a minute. It is a matter of experience to realize if the effect is physiological or due to seeing. The limiting magnitude of stars seen with a certain aperture depends on the specific fraction. For an 8-in. telescope it is: 14.2 mag (98%), 15.2 mag (50%), and 16.2 mag (10%). At a level of 10% the star flashes into view only for a few seconds in a minute. To increase the time of positive perception, the eye must be holding steady in the correct position over a long period. This is a real challenge and can be practised only by trained observers. Some even breathe more deeply or frequent than normal, to get more oxygen, but the amount of real success is doubtful. The normal observer can attain the 50% level of detection fairly easily. You can test your limit by using standard areas with known visual magnitudes. An example of a well-prepared testing ground is the open cluster M 67 [114] (Fig. 5.2); magnitude sequences for other clusters can be found in Luginbuhl & Skiff.

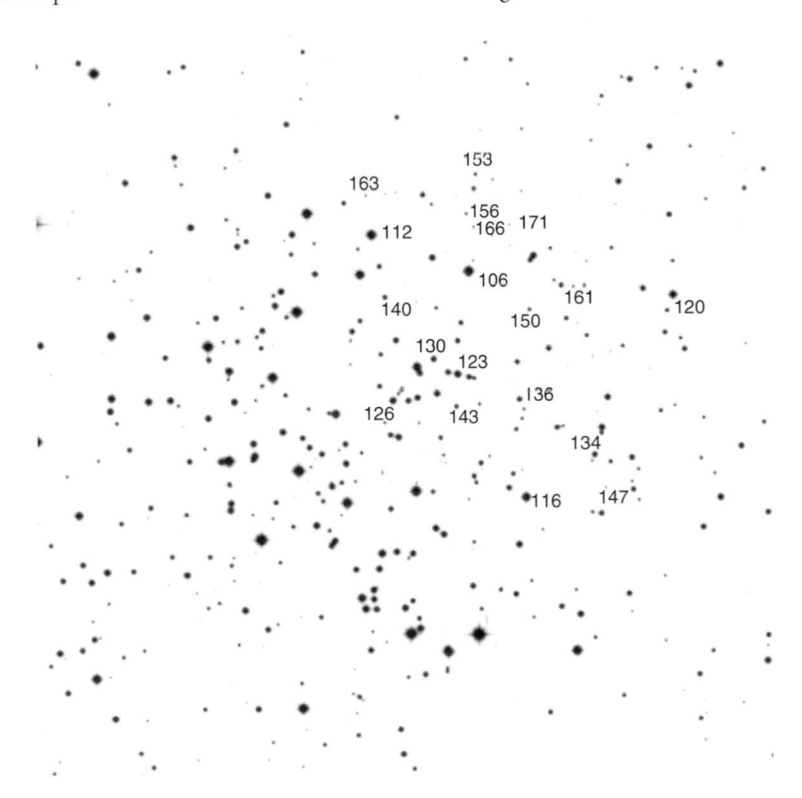

Fig. 5.2. Visual stellar magnitudes in M 67 (decimal point omitted)

The eye is also very sensible to subtle brightness variations and motions. The latter is used in a technique called "field sweeping" to recognize faint objects. A moving source is easier to detect than a static one (e.g., a faint satellite, entering the eyepiece, will be recognized immediately). In case of galaxies, we must create motion: try jiggling the telescope a very small amount while viewing through the eyepiece (not too much – don't loose the field). The direction of motion should consider the orientation of the object in the field. Take a faint edge-on or even superthin galaxy for instance. First, be sure it is in the field (by locating the relevant stars in the finding chart) and then check the orientation. Choose a higher magnification to increase the apparent size of the object. The best direction of sweeping is orthogonal to the thin line. Now you might see it!

Brightness variations can occur for quasars or AGN. Sometimes the change is not smooth, but perhaps the observer might luck out and catch a burst. Your eye is well prepared even for small amplitudes, even if the variation is fast. Perhaps a future visual observer will experience the beginnings of a Gamma Ray Burst!

Entrance and Exit Pupil, Perception

After explaining the physiological aspects, a bit more theory is needed to understand the effect of intensity, contrast, aperture and magnification on visual observing. All three major components, object, instrument, and eye, will be brought together now. Let's start with the concepts of entrance and exit pupil, needed to understand how bright extended objects, like galaxies, appear. Considering the sky background brightness, we are led to the quantity "contrast." Finally we will see, how magnification works – both for extended and point sources. Note, that we talk mostly about faint sources here, thus rods play the leading role – and averted vision is the relevant viewing technique.

Entrance and Exit Pupil

Entrance and exit pupil play a fundamental role for deep sky observing. In case of a telescope the entrance pupil is equal to aperture. In this chapter, the eye is the telescope. The entrance pupil is the diaphragm in the iris, which can change its diameter (denoted P). It normally protects us against bright light, which can overstrain the detectors on the retina. In case of darkness, the eye opening can reach a "maximum entrance pupil" (P_{max}) of 6–8 mm, depending on the physiological condition (young observers are favored).

The exit pupil measures the amount of light coming from the telescope. As already explained, its size (denoted p) depends on instrumental quantities only (aperture ratio, focal length of the eyepiece). The values of p and P are independent, which leads to three different cases. If the exit pupil is smaller then the entrance pupil ($p < P$), then all of the light enters the eye – but it could receive more. For $p > P$ only the fraction of light that fits into the entrance pupil is detected, all other rays are lost. Thus is $p = P$ perfect? Concerning the maximum light capture, the answer is "yes," but two things are necessary. First, we have to wait for maximum dark adaptation, to achieve, say $P_{max} = 7$ mm, then we must choose an eyepiece with guarantees the corresponding exit pupil of $p = 7$ mm at our telescope (e.g., $f' = 35$ mm for a 1:5 Newtonian).

You can of course use a larger exit pupil ($p > P$) for specific purposes, e.g., to get a larger field of view. However you then accept severe vignetting (i.e., light loss). Some observers don't like the abrupt vignetting that occurs when the exit pupil matches the eye opening

($p = P$). A very small movement of the head causes an immediate vignetting to occur. Making the exit pupil slightly smaller than the entrance pupil by choosing a different eyepiece can prevent this. Further benefits of $p < P$ will be discussed later.

Extended Objects

Is there an influence of p and P on the perception of objects? Yes, but we must distinguish between extended and point sources. An extended object (galaxy) appears as a luminous area. The ratio of exit and entrance pupil determines the brightness in which the object appears in the eye – regardless of its (nominal) surface brightness! The detected apparent brightness (denoted A) is given by $A = (p/P)^2$.

To understand the consequences of this relation, we will start with naked-eye viewing. Take the Andromeda Nebula as target, which is an oval patch of 1° (or even more) under dark sky conditions. Lacking a telescope and an eyepiece, what is p in this case? Without a limiting tube, the whole sky is the exit pupil! But whatever we estimate for p, the light will be cut by the eye's entrance pupil P. Thus we cannot receive more light than given by $p = P$, and consequently $A = 1$ is the maximum apparent brightness detected with the naked eye. Now use a telescope. As $p > P$ leads to light loss, we naturally choose an eyepiece giving $p \leq P$. Thus the apparent brightness of the object, as detected by the eye, is $A \leq 1$.

What follows sounds curious: Assuming a constant entrance pupil (constant dark adaptation), a luminous area appears in a telescope in no case brighter than with the naked eye. Imagine a trip to Mauna Kea. Our dark adaptation is maximum ($P = 7$ mm) and we have the benefit to observe the Andromeda Nebula through the Keck 10 m-reflector, using an eyepiece which gives $p = 3.5$ mm (most likely there is none). The apparent brightness is $A = (3.5/7)^2 = 1/4$, which is four times fainter than with the naked eye!

Contrast and Magnification

Theoretically your telescope would be limited to the Andromeda Nebula, M 33 and a few other targets – a miserable prospect! Does this meet your experience? Clearly this is not the case. Already with a 20 cm telescope (much less then 10 m) many galaxies can be seen in the eyepiece as extended objects, which are invisible to the naked eye! What is wrong? Obviously some important factors were ignored: mainly the night sky brightness, contrast and magnification.

The quantity A determines the apparent brightness of extended objects, but what is really important for visibility, is the brightness of the sky background – and A must be applied to this part in the eyepiece in the same manner. A maximal value of A guarantees a bright object, but unfortunately a bright background too!

A "typical" dark night sky shows a visual surface brightness of $V' = 13.1$ mag/arcmin^2 (which is equal to $V' = 22.0$ mag/arcssec^2); in the blue it is still a magnitude fainter ($B' =$ 14.1 mag/arcmin^2). Note, that this is not "dark" in the very meaning of the word. For demonstration, stretch your hand out against the sky – it will appear much darker! The eye can even detect much lower brightness levels, than presented by the "dark" night sky. Take your hand again. Illuminated by the night sky, you will see structural details even though only being lit by a mere 25% reflected light (thus 1.5 mag fainter).

Physically "contrast" is a signal-to-noise intensity ratio. Note that while intensity is an additive quantity, brightness is defined by its logarithm (thus magnitudes cannot be simply added). Let I_N equal the night sky intensity ("noise") and I_S the object's intensity ("signal"). Unfortunately I_S cannot be measured directly, we always get sum $I = I_S + I_N$ (as both intensities reach the detector at once). The object's intensity then results by subtracting the night sky from the detected quantity: $I_S = I - I_N$. This fact must be taken into account when reading values like $V' = 14.5$ mag/arcmin^2 for galaxies, which is 1.6 mag/arcmin^2 fainter than the night sky! How is this possible? As described, the observed surface brightness is the "sum" of 13.1 mag/arcmin^2 (night sky) and 14.5 mag/arcmin^2 (galaxy). After conversion to intensities this gives 12.8 mag/arcmin^2, which is (and must be) brighter than the night sky.

The eye is very effective in detecting faint objects under low contrast conditions. Contrast is defined by $C = I_S/I_N$. A high value is necessary for visibility, but not sufficient. The very quantity is called "contrast reserve" ΔC. It is the difference between the contrast due to the object (C) and a "threshold contrast" (C_T), which is the minimum contrast, needed for the eye to perceive a luminous area under the given sky conditions.

Now magnification enters the scene. What happens with a large, faint (i.e., low surface brightness) galaxy at higher magnification? Following the rules described above, the exit pupil decreases and thus the apparent brightness (A) of both the object and the background gets lower. Therefore the contrast remains constant. We might have won nothing – in theory. Fortunately, this is (again) not the whole story.

Remember, that the eye rewards a higher magnification in case of averted vision! Not only the amount of light detected by a single rod is important, but also the number of rods involved, i.e., the corresponding area of the retina covered by the light. With the aid of the ganglion cells, the eye–brain system is able to combine many rods to intensify the signal. Thus the perception depends on the "viewing angle" under which the object appears at the retina. Ideally this angle is 1°–2°. Most objects are not that large. For all smaller ones, simply increasing magnification will make it! Take for instance a faint detail in a galaxy, measuring 1′ on the sphere. A magnification of 60–120× is sufficient to blow it up to the required apparent size. Increasing the magnitude, some parts of the galaxy disappear, while other come out of the dark.

Not only the object's area in the eyepiece is important, but also the rest, i.e., the background. Its detection also depends on the area ratio. If the magnification is too low (small object, large background), the resulting area ratio ("signal difference") is insufficient for the brain. The object is lost in the background noise. A higher magnification dims both the object and background (constant contrast), but the ratio of their individual sizes on the retina increases, the object appears. If magnification gets too high, the object fills most of the field of view and the signal difference decreases again. Thus there must be an "optimum detection magnification" (ODM) for extended objects.

Concerning the ODM we need to distinguish between two cases, one of which finally introduces the quantity "aperture." For faint, small galaxies (moderate surface brightness) the ODM is high, thus we need a sufficient aperture. For faint, large galaxies (low surface brightness) the ODM is lower. We don't need large telescopes in this case! Thus a small telescope can readily detect large low surface brightness galaxies of the Local Group, while a large aperture often reveals nothing.

To calculate the contrast difference (ΔC) and the ODM the following quantities must be known: surface brightness of the night sky and the object (nominal value), and the telescope aperture. A positive value of ΔC promises visibility. A zero or negative value means simply: "next target!" Mel Bartels has developed a nice tool to calculate the relevant quantities [115,116]. It demonstrates impressively, that in most cases the darkness of the night sky is more important than aperture. Try for instance the low surface brightness galaxy IC 2574 (Fig. 5.3).

Fig. 5.3. A difficult object: IC 2574 ("Coddington Nebula"), a member of the M 81 group in Ursa Major

Point Sources

For point sources like quasars things are different (Fig. 5.4). If we use an exit pupil of $p = 1$ mm in case of maximum dark adaptation ($P = 7$ mm), we get an apparent brightness of $A = (1/7)^2 = 1/50$ for the sky background. This is 50 times darker than it appears to the naked eye. Any extended object would be equally dimmed (constant contrast), but not a quasar. Thus the contrast increases and the quasar stands out clearly in the dark (by the same "trick" we are able to see stars at daylight in a telescope). This implies a high magnification. Take for instance a 50 cm Dobsonian: we get $m = D/\mathrm{p} = 500$ mm/1 mm = 500×. Choosing a much lower magnification, let's say $m = 100\times$, we would get $p = 5$ mm, and thus $A = (5/7)^2 = 1/2$. This might be well for an extended object. Compared to 500×, the background appears 25 times brighter now: the quasar will be lost in the noise! Moreover a brighter background will influence the dark adaptation.

We have already discussed the limiting visual magnitude in the chapter on "averted vision" is influenced by the observers experience and techniques. We may only add here the role of magnification, as one of three quantities, which determine the theoretical visual limiting magnitude of point sources (as described in [117]). The other two are aperture and the naked-eye limiting magnitude at the zenit (to be discussed in the following chapter). They quantify the factors background sky brightness, light collection, and transparency, respectively.

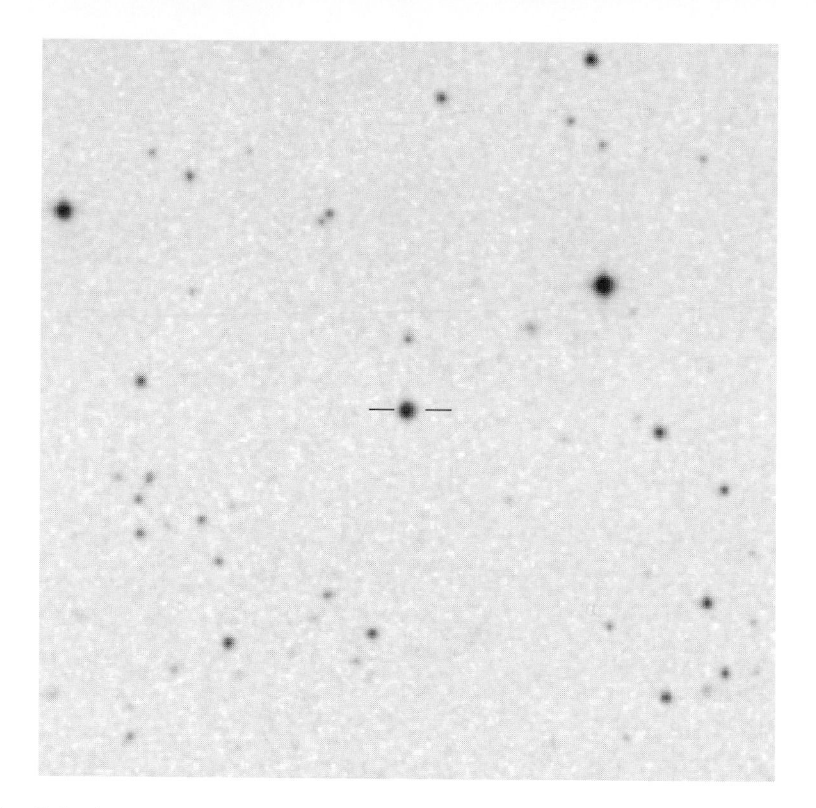

Fig. 5.4. A point source: PG 1634+706, an extremely luminous quasar in Ursa Minor

What happens at high magnification? The limiting magnitude will reach a certain degree of saturation. If the seeing is good enough, then the object looses its point like character by becoming a tiny disk, due to diffraction. Any further magnification will affect its brightness (parallel to the sky background), thus the contrast becomes constant. This effect is essential below an exit pupil of $p = 0.7$ mm, which is equivalent to a magnification of 714× in case of a 50 cm Dobsonian.

Atmospheric Conditions, Observing Site

Visual observing depends crucially on the condition of the atmosphere. Two quantities that act on all kinds of objects are relevant: seeing and transparency. Seeing reflects the turbulence of the atmosphere and transparency its lack of particles – relevant for air and light pollution. Seeing influences telescopic resolution, while transparency affects the apparent brightness and contrast. To observe fine details in a bright galaxy good seeing is necessary, whereas good transparency is essential for low surface brightness galaxies.

The documentation of atmospheric conditions can refer to short or long time scales. The latter is essential to find an ideal observing site, e.g., to build an observatory. "Short time scale" means checking the conditions during the observing session. It is important to note seeing and transparency in the observing log. This characterizes the value of the observational results and is important for comparison.

Level	Seeing	Image
I	Perfect	Calm image, without a quiver
II	Good	Calm image over many seconds, interrupted by slight quivering
III	Moderate	Larger air tremors that blur the image
IV	Poor	Constant troublesome undulations of the image
V	Very bad	Hardly stable enough to allow a rough sketch to be made

Table 5.1. Antoniadi's seeing scale

Seeing

Turbulence is mainly due to convection (vertical flow) or advection (horizontal flow) of the air. These motions are driven by temperature or pressure differences. Local variations of air density influence the diffraction conditions that affect light rays in many ways. The net effect is called "seeing," and is visible to the naked eye as rapid changes of the brightness (twinkling star), position (scintillation), and color.

Seeing affects the sharpness of the image, which is most important for small or stellar objects like quasars or compact galaxies. Galaxies are smeared out; details like knots or even spiral arms vanish. In comparison with transparency, seeing has less influence on brightness and contrast.

There are various scales to quantify seeing [104]. Absolute scales, which are difficult to use for the amateur, measure the diameter (FWHM) of the "seeing disk" in arc seconds. A bit more subjective, but still difficult to apply, is the Pickering scale defining 10 levels. It uses the diffraction disk ("Airy disk") and rings of an in-focus star, seen with a standard 5-in. refractor. A simple, but pretty subjective scale was introduced by Antoniadi (Table 5.1). It defines five levels, depending on the appearance of a star in the eyepiece. The aperture and magnification should be noted.

Transparency, Faintest Star

A common measure of transparency is the naked-eye magnitude limit, called "faintest star" (fst). The transparency varies over the sky, e.g., fst-values taken at the zenit (ZLM = "zenital limiting magnitude") or at low elevation are significantly different due to light pollution. Thus it is useful to determine the limiting magnitude near the observed object. But this does require detailed charts. A sky mapping software is helpful, but you must check if the stellar magnitudes are truly visual. Fortunately there are some standard areas that are spread over the sky. The "north pole sequence" is a classic example, easily visible for northern hemisphere observers. In the course of time one gets familiar with the area and the determination of "fst" doesn't take much time.

Be sure, that there has been enough time for good dark adaptation. It is better to place the determination of the visual magnitude limit in the middle or at the end of the observing session. Averted vision is generally accepted, but the star must be held in the eye, a mere "flash into view" is not sufficient. Thus the fst-value is pretty subjective. Those wearing glasses have problems to yield a reliable result.

A similar method is based on star counts in a given area, e.g., the great square of Pegasus. The transparency is roughly proportional to the number of visible stars. Another technique is to look at the naked-eye appearance of the Milky Way. Note, that it is quite absent from the northern sky in spring. John Bortle has created a scale with nine levels

Table 5.2. Bortle's transparency scale

Bortle	Sky	Description	fst (mag)
1	Excellent dark	Zodiacal light, gegenschein and zodiacal band are visible; M 33 is a conspicuous object; the bright Milky Way clouds cast obvious shadows	7.8
2	Truly dark	No stray light; M 33 is easy visible; the Milky is highly structured	7.3
3	Rural	Weak light pollution near the horizon; the Milky Way looks complex; M 33 is easily visible by averted vision	6.8
4	Rural/Suburban	Fairly obvious light domes over population centers; zodiacal light is still visible; the Milky Way is impressive, showing only the most obvious structures; M 33 is barely visible by averted vision	6.3
5	Suburban	Light pollution is present in any horizontal direction; zodiacal light is very weak; the Milky Way is dim near the horizon and nearly washed out at the zenit	5.8
6	Bright suburban	Up to an altitude of 35° the sky looks gray; the Milky Way is only visible overhead; M 31 is only modestly apparent	5.3
7	Suburban/urban	The sky is entirely covered by diffuse gray light; the Milky Way is nearly invisible; M 31 is a difficult object	4.8
8	City	The entire sky appears in white–gray or orange; a newspaper can be read easily; M 31 is extremely difficult	4.3
9	Inner-city	Bright sky, wherever you look; the Pleiades is the only deep sky object; only the brightest stars or constellations are visible	≤3.8

(Table 5.2), describing the visibility of the Milky Way and prominent objects like M 31 and M 33 under different sky transparency conditions [118]. The first level of the "Bortle scale" is exceptionally rare, but the mid-range (4–6) is quite frequent. Each level can be associated with a certain fst-value (in 0.5 mag steps).

The Bortle scale can also be (roughly) quantified by the surface brightness of the sky background. This value is needed to calculate the ODM, as discussed above. An excellent/truly dark sky (Bortle 1/2) measures 13.1 mag/arcmin^2 or better, but at an urban site (Bortle 7 or 8) it reaches only 12.1 mag/arcmin^2. This drastically limits the visibility of low surface brightness objects – as a matter of contrast reserve (ΔC). A rule of thumb says, that galaxies with $V' = 15$ mag/arcmin^2 are at the limit for medium-sized amateur telescopes, even under a sky with Bortle 2 or better. Objects with $V' = 13.5$ mag/arcmin^2 are even pretty difficult at Bortle 5 or 6, e.g., M 31 with $V' = 13.4$ mag/arcmin^2. Obviously the integrated magnitude, which is $V = 3.5$ mag, is less important. For M 33 with $V' = 14.1$ mag/arcmin^2 (but $V = 5.7$ mag) an even darker (rural) site of Bortle 4 or lower (which gives around $V' = 12.7$ mag/arcmin^2) is needed. M 81 can be seen with naked eye only with Bortle 1 or 2 (Fig. 5.5).

Observing Site

A really dark site is more precious than a large telescope. The only advantage of large aperture at a poor site is to achieve high magnification, dimming the background.

Fig. 5.5. One of the most difficult naked-eye objects: M 81 in Ursa Major

The best places are characterized by: no stray light and air pollution, calm and dry atmosphere, many clear nights, not too far away, easy to access, save. No question, such conditions are hard to find! The reality is much poorer.

If you own a garden observatory or observe from your balcony, you have no choice concerning light pollution. Try to shield any stray light as good as possible. Even for light polluted sites, there are plenty of celestial treasures to observe [119]. Such a set-up has a major advantage: short reaction time. Whenever the sky clears up, you are ready to observe. This is important if your task is monitoring light variations of AGN, quasars or BL Lacertae objects – perhaps you will discover a burst?

If you and your telescope are mobile then try to find the best site, but the drive should not take more than 1 hour. A remote site may be a nice place, but how to get there safely in the dark? Ask for permission if necessary. Try to omit any unwanted surprises at night. To get information about the observational qualities, some on-site testing is needed. It is pretty easy to recognize the lack of light pollution, but the evaluation of the site-specific atmospheric conditions is not straightforward. Inspect the terrain during the daylight too. What's about the ground? Maybe there is concrete or asphalt to put up the telescope. Note, that such material can get very hot at daytime, radiating the heat after sunset. Thus you must reckon with convection, which can lead to bad seeing in the first half of the night. The same is true if walls are nearby. Mountain forests store heat too, which negatively influences the atmosphere during the night. Any wind blowing over the trees will produce a turbulent air mixture. Thus it is better to keep away from forested areas and instead look for large, relatively flat rural areas with an unobscured sky view. Omit humid places, e.g., near creeks and small bodies of water, which may lead to excessive dew. Hillsides can be subject to cold airflows. If conditions are turbulent, this will influence seeing too. High places are often above the inversion layer in autumn and winter, which guarantees a relatively warm, dark and dry night.

Even better are islands with high mountains, producing a stable inversion over long periods. The air is calm and dry, and coupled with laminar air flow leads to excellent seeing. Top sites are La Palma or Hawaii. But high elevations can cause a serious problem: low oxygen levels. Your eyes are, in particular, sensitive to changes in oxygen levels, which can lead to a dramatic reduction in the degree of dark adaptation. Nor is it pleasant to observe with headache – so its best to have several days to adjust to local conditions. Americans, especially those living in the southwestern states, have the opportunity to observe from excellent, high elevation desert sites. Here the Milky Way appears so bright that when it rises above the eastern horizon "first timers" often think it that the skies are becoming "cloudy."

Chapter 6

Observing, Recording, & Processing

Any observing session needs pre- and post-processing, which covers many different activities [120]. One must compile an observing program, choose the right clothing, and check the equipment. Conscientious preparation is an important factor. Missing an eyepiece, filter or an atlas on an excellent night can be disheartening.

The central activity is of course observing. We have already discussed the physiological and physical aspects. In this chapter we focus on finding the objects, featuring "starhopping," and recording the results. The latter includes textual descriptions, sketches and drawings. Finally one can analyze and publish the observations.

The Observing Session, Starhopping

The Right Clothing

Most observers do not have the benefit to practise their hobby in a warm climate. But even in a desert it cools quickly. Wind and coldness ("wind-chill" effect) are unpleasant factors for stargazers and may influence concentration and patience. Who only invests in the instrument, but not in the right clothing, cannot observe successfully. At night the mind and body are most sensitive – so all kinds of stress should be avoided.

Effective protection requires three-layer clothing: The inner layer (underwear) should be made of cotton or synthetic microfiber. This absorbs sweat; otherwise evaporation would cool the skin too much. The middle layer is made of heat insulation material, e.g., eiderdown, wool or fleece, to keep the body-heat like a cocoon. The outer layer must transmit water vapor to the environment and keeps all wind away.

Be particularly careful with head, hands and feet. Fleece bonnets are good for protecting the head, where 30% of the heat is lost. Gloves are a bit problematic. The fingers are used at the telescope, thus thick gloves would hamper their mobility. Wearing the right footgear is most important. Cold feet can be the death of any observing session. The sole must be thick enough for heat insulation. Be careful with the socks: leave space for the feet to move. If wedged, they cool out rapidly. The muscles must work for heat production and a sufficient air layer will protect against heat loss. Moonboots or light snow boots are always a good choice.

Preparing the Equipment, Useful Tips

Not only the body should be protected, but the equipment too. Optical surfaces must be kept clean. One should be prepared to make small repairs if necessary (toolbox). Put batteries on charge or keep new ones (e.g., for the red light) on hand. Always bring food and drinks with you, and warm beverages are especially welcome on bitter nights.

Another crucial aspect to a successful observing run is the transport of equipment. Small items like eyepieces, filters, etc. can be stored in an equipment case or photo bag. Compact telescopes like SCTs are delivered in a case or stout box(es) that are very useful for transport. But in case of Dobsonians, things are a bit more difficult. A solid tube can be wrapped in a thick blanket, while truss-tube models have many different parts that must be stored safely. Be particularly careful with the primary and secondary mirrors as they can be easily damaged. Often the secondary is mounted in a baffle, which must be carefully covered.

If you can arrange it, try to assemble the telescope at dawn or even earlier. Keep it away from the street to avoid turbulence and headlight of approaching cars – they will destroy your dark adoption in a second. Check the collimation, align the finderscope, put your eyepieces in order, prepare the red light and your charts. The hood of your car or the back latch of a pick-up truck are good places for your sky atlas and finding charts.

At dawn the transparency can be already derived from the colors of the sky (Fig. 6.1). The upcoming night will be exceptional, if the sky shines green in the west. You may notice dew at that time, which is a good sign too. While your telescope cools out and your eyes are adapting, you may have time to visit some familiar objects with a binocular. Star clusters are nice targets. This is also the time for a first seeing check. Defocus a bright star

Fig. 6.1. At dawn

to observe the quality of the diffraction image. Maybe there is still warm air in the tube and around the observing field. Prepare your observing program and get familiar with the relevant sky area of your first target. Always be flexible in your program and feel free to change it depending on the night to come.

Is it recommended to observe alone? Don't do so if you are not familiar with the area and mentally unstable – remote sites can be scary! Any kind of stress reduces the perception. Perhaps you have no choice, so music can help against loneliness and fear. In group settings it can be more pleasant – but sometimes more distracting. It is always valuable to communicate to others about your visual impressions or perhaps get ideas for new targets. Anyway, be prepared for a case of emergency, and keep in mind where a phone is located.

It's advisable to limit your intake of nicotine and alcohol since it has been shown that these drugs will have a deleterious effect on visual acuity and dark adaptation. It is also important to be relaxed at the telescope. A cramped position affects your concentration. Use an adjustable chair to relieve stress.

Starhopping

Now you're ready to find objects. We ignore any sophisticated tools, like digital setting circles and GoTo, and instead rely on the simple but effective technique of "starhopping" [121,122]. This means jumping from star to star and field to field on a defined path to finally reach the target. This method must be learned and perhaps the best way is using a pair of binoculars. The transition from naked-eye viewing immediately leads to the problem of orientation. In the binoculars, the number, brightness, and the apparent distance between stars appear different. The situation can be even worse in a finderscope especially if the view is inverted and mirror-reversed. This can be a real challenge for any beginner!

Let's now push things to the limit and try to find a faint galaxy or quasar. Step by step in an "incremental" sequence going from the naked eye – Telrad – finderscope – low power eyepiece at the telescope – high power eyepiece. Thus starhopping is normally "three-dimensional"; that is by moving along a (two-dimensional) path on the celestial sphere, while getting deeper into space at the same time by using a larger instrument (higher magnification). By changing the instrument, the size and orientation of the field will also differ. You first start with the Telrad, showing an upright image. Next you look through the finderscope with a star diagonal, which is an inversed and mirror-reversed field. Finally you use a Newtonian, where the image is not mirror-reversed but rotated, depending on the position of the focuser. Your brain has a lot to do.

Any starhopping sequence needs preparation in the form of suitable charts: planisphere for the naked eye or Telrad, a small scale star atlas (e.g., Sky Atlas 2000.0) for the finderscope, a large scale star atlas (e.g., Uranometria) for the low power eyepiece, and finally a printed finding chart, showing the target from a sky mapping software or DSS image (e.g., Guide, RealSky) for the high power eyepiece. Starhopping works by using star patterns ("mini-constellations") and connecting lines on any scale. Reference stars of appropriate brightness are then used to connect the sequential star fields. Avoid using steps that are too large in case of difficult targets or small telescopic fields.

Sort your charts and place them on a solid surface, e.g., a camping table or the hood of the car. Use some clothespins or large clips to fix them in case of wind. Clipboards for holding charts are very useful when on top of a ladder (often the case with large Dobsonians) or if you are a good distance from your chart area.

With growing experience any starhopping will be successful – which only means, that the final field is found and centered in the eyepiece. The object may still be invisible! Now you're in the situation to apply various viewing techniques, like averted vision or field sweeping, or to try a filter. If you are successful, a lasting feeling of accomplishment is guaranteed! Regardless how faint the object is, take time to observe it, e.g., 10–15 min for a simple galaxy, 30–60 min for a complex one, like M 51. The more time we spend observing, the more that will be seen.

By repeated starhopping to the same target, the brain learns to find the way without any chart. This is both valuable and impressive when presenting the sky to beginners. Starhopping opens the door to the "personal" sky – this is observing at its best. But a starhopping tour is not a fixed road. There are alternative paths and many turn-offs leading to interesting objects nearby.

Example: A Starhop to 3C 66A

As an example we will hop to the BL Lacertae object 3C 66A in Andromeda (Table 6.1). The best observing season is autumn, when the constellation is near the zenit on the northern hemisphere. For medium transparency (Bortle 3–4) the 15 mag stellar object should be visible in a 12-in. telescope. Sometimes even 8–10 in. will make it, as 3C 66A is variable and can reach 14 mag (Fig. 6.2).

A good starting point is γ And, the easternmost star in the Andromeda chain (Fig. 6.3; *Guide 8* plot). Point your telescope (via Telrad) to the star and center it in the cross hair of the finderscope. In the main instrument you will see a colorful double. With the naked eye fix a straight line to Algol = β Per (which is to the east) for further orientation. Using the finderscope (Fig. 6.4; *Guide 8* plot), you will meet an easily visible trapezoid extending 1.5°, or around one-third the distance. The large edge-on galaxy NGC 891 lies at the upper edge of this line tilted about 60° to the left. Three stars are of equal brightness (around 7 mag); the lower star is a magnitude brighter, with an optical companion north. Fix a line of the two northern stars and extend it straight to the east by a similar distance. You will see a faint star of 8.5 mag, designated SAO 37990. If you don't see the star, simply center the cross hair to an imaginary point there. The star is plotted in the Uranometria 2000.0, which shows 3C 66A too. This suggests that the BL Lac object is an easy-to-find target, but in reality the stellar object is more than 5 mag below the magnitude limit of the atlas.

The star should now appear in the low power eyepiece at the telescope. For the last step, a finding chart is necessary. It can be prepared with the DSS using *Guide* for identification (or a similar software with a quasar catalog). The star has three companions, all around 11.5 mag, forming an equilateral triangle to the south (diameter 2.5′). Connect

Table 6.1. Data of 3C 66A

Position (J2000.0)	02 22 39.6 +43 02 08 (And)
Type	BL Lac
Redshift (z)	0.444
Distance	1,456 Mpc
Apparent magnitude	14.0–15.8 mag
Absolute magnitude	–26.5 mag
Light travel time	3.9 Bill. years

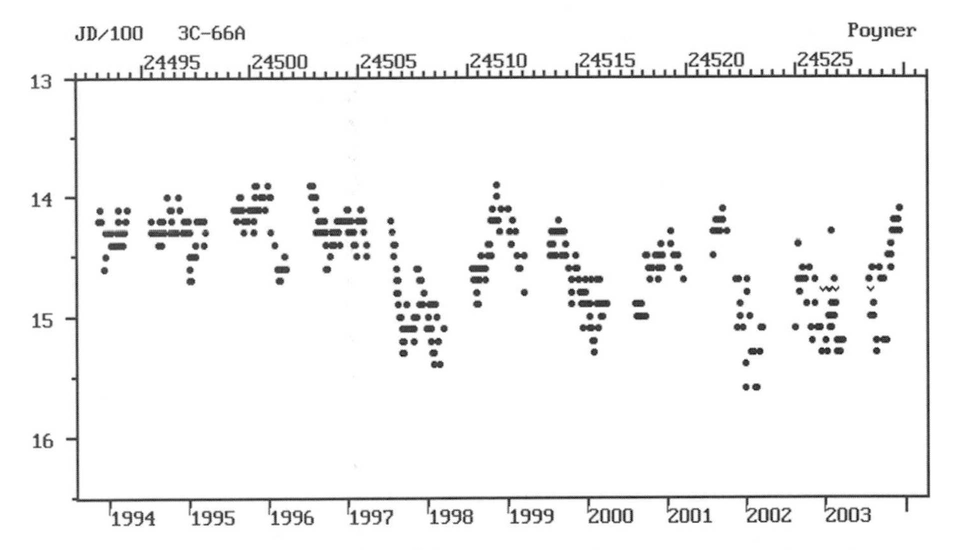

Fig. 6.2. Brightness variation of the BL Lacertae object 3C 66A in Andromeda

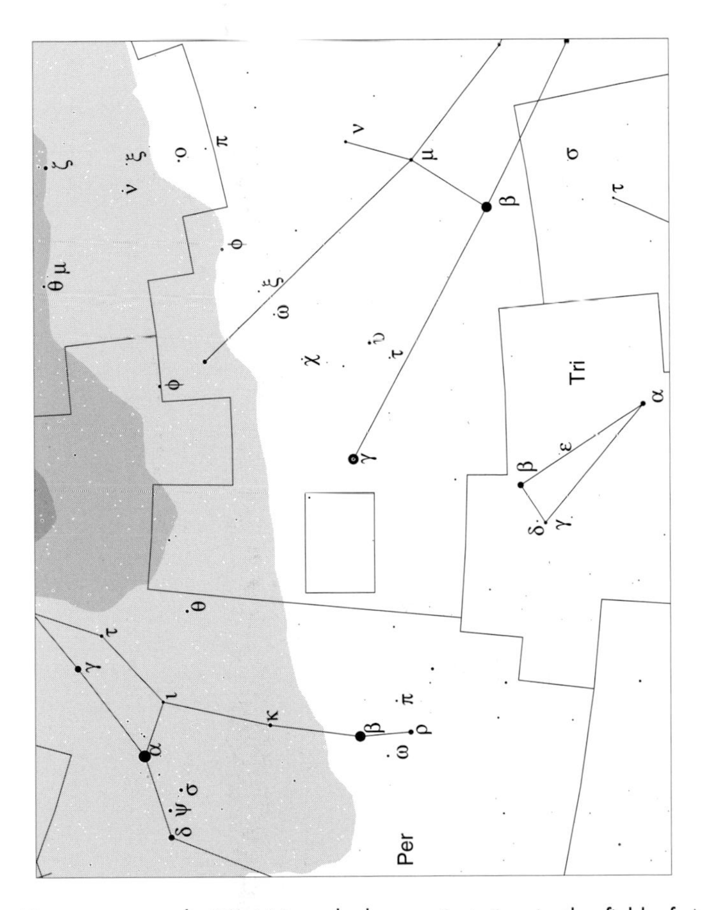

Fig. 6.3. First step towards 3C 66A: naked-eye orientation in the field of Andromeda and Perseus (see text)

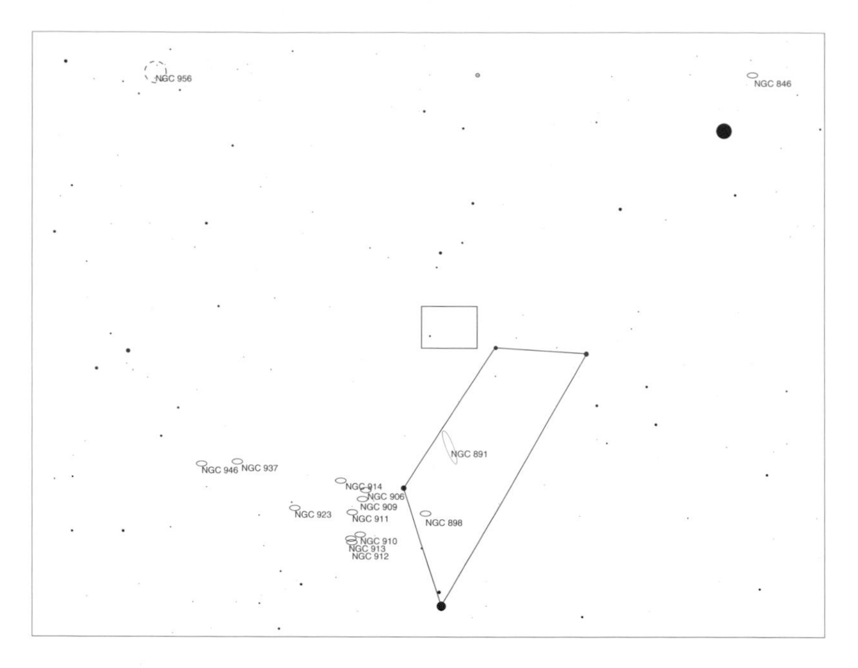

Fig. 6.4. Second step towards 3C 66A: finderscope orientation (see text)

the eastern star of the triangle with the SAO star and extend the line five times to the northwest. You will meet three faint stars in line (12.3–13.7 mag, extending 1.3′). You are now pretty near the quasar and can use a higher magnification. The three stars point directly to 3C 66A, which lies 2′ southeast, i.e., 1.5 times the extent of the triple (Fig. 6.5). If the object is not immediately present, try averted vision. Just imagine: those faint photons have travelled nearly 4 billion years to reach your eye (Table 6.1).

That's not the only object at this region. Maybe you already have noticed some faint patches in the field. We have landed in a (physical) group of three galaxies: UGC 1832, UGC 1837, UGC 1841. They are situated in the extreme foreground of 3C 66A at a distance of 90 Mpc and thus not related with the BL Lac object. It is interesting, that 3C 66A dominates a remote cluster of galaxies – an appropriate target for the HST. UGC 1841 is the brightest member of the galaxy group is identical with the radio source 3C 66B. Only 25″ below is the faint, extremely compact object V Zw 230, which is associated with the galaxy. Due to its variability this is a most interesting target for telescopes of 14 in. and up. The data of these galaxies are collected in Table 6.2.

The redshifts and their location indicates that all three galaxies are peripheral members of the galaxy cluster A 347, which has a mean radial velocity of 5,516 km second. The cluster, which is another interesting target, is located 60′ southeast. Still nearer is the large edge-on galaxy NGC 891, 40′ south (Fig. 6.6). It is not visible in the finderscope, thus center the position between the two trapezoid stars and use a low power eyepiece. Be

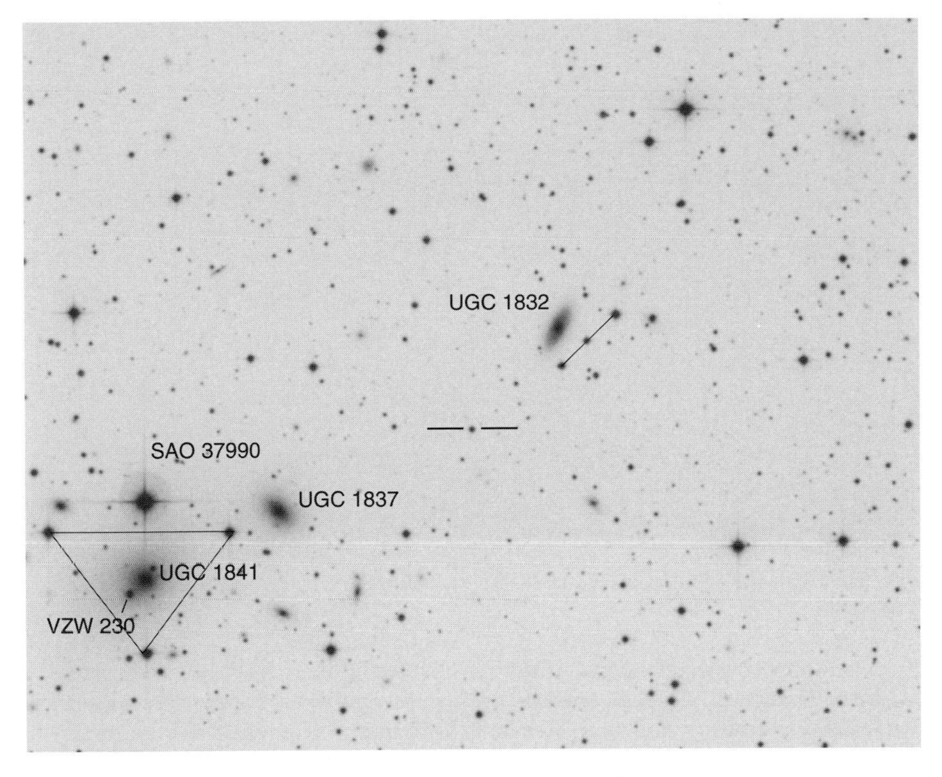

Fig. 6.5. Final step: field of 3C 66A in the telescope

Table 6.2. Galaxies near 3C 66A

Galaxy	UGC 1832	UGC 1837	UGC 1841	V Zw 230
Type	Sa	S0	E2	C
Brightness (*V* mag)	15.3	14.9	14.6	16.0 var
Size (arcmin)	1.1 × 0.7	1.2 × 0.9	3.9 × 3.0	0.15 × 0.15
Radial velocity (km/s)	5,913	6,582	6,373	6,595

prepared to see a 13′ long streak of faint light, cut by an absorption band (dark lane). An 8 in. should show this 10.9 mag galaxy well.

Observing Log

Try to keep an observing log. It is nice to read in case of bad weather – or even many years later. Keep in mind: What is not documented, regardless of the specific form, is forever lost! Those who are not highly gifted sketchers might favor a textual description. So how does one describe the visual impression of an object, say a bright galaxy? Words are nice, but any added rough sketch makes things much clearer.

Let's mention a critical theme concerning visual observations and photos (compare the remarks given at the beginning of Section III). As it often happens, an observer

Fig. 6.6. Bonus object near 3C 66A: the fascinating edge-on galaxy NGC 891 in Andromeda

"believes" to have seen "something." This may be the result of a bias in the form of images he has previously seen. Try to ignore these temptations and accept the truth: its no shame, if you have seen nothing! And it can even be valuable information for other observers.

Textual Description

All you need is pencil and paper. At the telescope only a scratchpad should be used, as the observing log can get wet. Don't write long treatises in the dark – instead try to use shortcuts, which should nevertheless be readable. At home you can paste or re-enter your writings (and sketches) into the log, but don't wait too long. Try to avoid changing the actual content of your notes.

Instead of writing notes at the telescope, you can use a dictaphone or mini tape recorder. It saves time, but needs preparation (batteries, tapes). Even more sophisticated is a digital observing log. You can store all your observational results on the computer, but avoid doing the job directly at the telescope. Even if the screen can be dimmed in red, it destroys some of your dark adaptation. To avoid this, pick up some rubylith film from an art supply store (or telescope vendor) to cover the screen. The same is true for using digital charts: try to print out them in advance and use a weak red light.
Any observing session, should be recorded by the following basic data:

- Date, starting/ending time
- Location, altitude
- Seeing (e.g., Antoniadi scale)
- Transparency (e.g., fst, Bortle scale)
- Type of telescope, aperture, focal length.

For an individual observation note the

- Time (perhaps specific seeing/transparency conditions),
- Magnification (exit pupil),

– Field of view, accessories (e.g., filter), and the
– Observing technique (e.g., direct/averted vision, field sweeping).

We now come to the description of the object and its surroundings. But first a few remarks on style in advance. Astronomy is an exact science, but this does not imply that a scientific language is required. Try to use graphic expressions, but don't exaggerate.

A galaxy looks different depending on magnification, field of view, filter used, etc. Thus it is difficult to estimate brightness and size. Try to compare with other objects in the field, but at the same time don't be ambiguous. For instance it makes little sense to note a galaxy to be left, right, up or down of a certain star. The brightness of galaxies can be estimated using reference stars with known magnitude, but it is generally difficult and incorrect to compare point and extended source.

To get orientation and position angles, the main directions (north, east) must be determined. You can practise the drift method. Without tracking, all objects move westwards. If a star is located to the west of a galaxy, it will precede; to the east it follows the galaxy. North is the direction to the pole, which is easily noticed by turning up the declination axis. For a Dobsonian one must move both axis.

To estimate the apparent size of a galaxy, it's best to go beyond using the terms like "large" or "small." You can quantify the larger and smaller diameter (in arcmin) by comparing them with the dimension of the field of view or the distance between two stars (which can be measured later using a chart). The form can be described as round, oval, elongated or even irregular.

There are standard shortcut descriptions in the literature, e.g., in the classic NGC. In case of NGC 660 (Fig. 6.7) one reads "pB, pL, E, bM, r," which means "pretty bright, pretty large, elongated, brighter middle, resolvable (mottled, not resolved)." In addition to brightness and size estimates, you can answer standard questions:

– Is the form round, oval, or irregular?
– Is the center bright, diffuse, compact or even stellar?

Fig. 6.7. The peculiar SBa galaxy NGC 660 in Pisces

– Are there any structures like spiral arms, dust bands or knots?
– Is the edge sharp or diffuse?
– Are there stars superimposing the galaxy?
– Are there other deep sky objects in the field?

There will be many objects, where most of these questions cannot be answered. Thus the observing log will show only a few notes in these cases. Don't add objects that are not actually present. Examples of good observation notes can be seen in the treasure of textual descriptions that are presented in the last section!

Sketches and Drawings

A sketch is always useful – and, if correct, may tell much more about the object and the surrounding than a mere text. No great talent is necessary, but there are some points to consider [123]. A sketch is not a drawing; it shows only the main features. Try to fix them already at the telescope. A small tray to fix the paper is useful. The red light can be already mounted there, or at a headband.

In previous sections, we have seen that good observation techniques will greatly enhance the visual experience. The same is true for sketching. Do not plot all details "at once," but step by step. It is recommended to use different magnifications to record the whole picture, but there should be one master field of view.

You can use a template (circle with orientation markers and scale bar) in which all information can be inserted. Start with the brightest stars, defining the frame for the object and nearby fainter stars. Mark wrong entries, but don't use an eraser, it could smudge the sketch. An alternate way is to print out the starfield (to a sufficient magnitude limit) with a sky mapping software in advance, omitting the object. This is useful for crowded fields. Mark stars which are not detected and add those not plotted (the data can contain errors). Continue with outlining the object's shape and add the details.

At home you should copy the sketch on clean paper, correct errors and enhance faint structures with a blending stump. This can result in a fine drawing, but this may require the development of skill and technique. Intensify only real structures; don't add any unseen things (known from photos). The most impressive looking drawings are on black paperboard with a white pencil or pastels. But your black on white sketch can be easily converted to this format. A sketch can be scanned, processed with standard software and finally inverted. The result looks like a photo – except this is a realistic rendition of the image at the eyepiece. You can then transfer them into a computer log, or transfer the sketches into the observation log. It is often more convenient to put them in plastic envelopes in a separate folder or just maintain separate sketchbooks. Just pick a method that works best with your style of observing and record keeping.

Analysis, Evaluation, and Publication

Your treasured collection of results (descriptions, sketches) will grow with time, but what to do with them? Maybe you have already discussed your observations with other amateurs. But you can do more. Analyze and evaluate your data, using the advantages of spreadsheet software. Input all relevant data in the table: object, type, appearance (e.g., size and brightness estimates), place and time of observation, atmospheric conditions, aperture, magnification, etc. It is useful to add catalog data: coordinates, brightness, size, cross-references, literature, etc.

The resulting database can be sorted in many ways and is subject to subsequent statistical analysis. Not only mere counts are interesting – but the type of objects observed or even trends in the climatic data. For example, you might be curious on how many edge-on's were observed over the past year. Try to analyze your data in relation to quantity or qualitative features. A few examples:

- Brightness variations of a variable galaxy/quasar (see for example Fig. 6.2)
- Visibility of edge-on galaxies under different conditions
- Appearance of extragalactic globular clusters
- Sizes of face-on galaxies
- Exploration of the Hubble classification sequence
- Cases of galaxies with superimposed stars (to avoid "supernova" sightings; Fig. 6.8)
- Appearance of galaxies in the zone of avoidance
- Comparison of visual vs. photographic magnitudes.

How about a publication in a magazine or on a website? Many observers think that their results are not "scientific" enough. Some lack the courage to publish, or are loners who would rather ignore the rest of the (observing) world. We will never know anything about their accomplishments.

What are the quality requirements for a magazine article? This depends on where to publish. If your target is one of the professional journals like *Astronomical Journal* or *Monthly Notices of the Royal Astronomical Society* it is nearly impossible to place your

Fig. 6.8. M 108 in Ursa Major: prominent galaxy with superimposed star

work there. Fortunately there are many lower levels, but even popular magazines like *Sky & Telescope* or *Astronomy* are not easy to satisfy – and the process of publication can take some time. For beginners it is recommended to publish in online-magazines or journals of astronomy clubs. Your contribution is most welcome there! No courage is required – and you won't have to worry if it's "scientific" enough.

It is not always necessary to present observational challenges. Prominent targets are always interesting – and there are always readers, for which the object is still new. For an owner of a 4 in. reflector it is important to read about how many Messier galaxies can be seen, or if NGC 604 in M 33 is possible. What looks redundant for a high-end observer will be a challenge for the beginner. Anyway, tell the public about your experiences – even negative results can be enlightening.

Try to investigate your subject a bit before writing – and you might be the one to profit most. Be critical with your text, drawings or photos. Your results look more reliable if you discuss possible errors or problems. Try to give an outlook on follow-up observations, which may generate response.

Finally, present your results on a star party; some offer a lecture program. Communicate with experienced observers there – it will be very informative and inspiring. But don't forget observing!

Section III

What to Observe? – The Objects

You might expect a single, long list of galaxies in this section. But a huge dataset, like the *Deep Sky Field Guide*, may eventually turn your enthusiasm into resignation. What to do, if your sky mapping software plots thousands of anonymous galaxies on the screen? In planning an observing session, it might be more adequate to collect and sort the targets under various themes or topics. Therefore different categories, each with a representative selection of objects, are presented here. To get an impression how they look, their appearance in different apertures is also described. This can be helpful to select and grade similar cases, which meet your specific requirements and preferences. Additional notes are also provided for selected objects that have interesting astrophysical properties, unusual features and/or background history. Then use this information and your imagination to create your own personalized observing programs.

What was the reason to choose a combination of photo/textual descriptions in this book? It is generally known that visual observations and photos of galaxies are rather difficult to compare. Conventional photography or CCD imaging crucially depend on technique (spectral sensitivity, filters, image processing) and even images of the same object can look quite different. But in a sense they are "objective." Against that, the visual impression is naturally subjective. Nevertheless the eye is able to perceive fine structures and contrast differences, even at low light levels – and works occasionally better than artificial detectors.

Textual descriptions can be useful if they concentrate on the main observational facts, while noting any kind of uncertainty. What about sketches? Based on the visual impression, the degree of subjectivity is equal. But the medium "paper" implies a special aspect: the danger off adding "virtual" features. Uncertain structures are difficult to assign. Thus a sketch or drawing appears like a "fact" – and gets a rating comparable to a photo. Only experienced observers can judge its value. Thus photos are preferred, which show the physical nature of the object best (remember the book is not a mere observing guide).

For each individual galaxy or higher order system (e.g., group, cluster), presented in this section, the best available data are given. The morphological types usually refer to the Hubble classification, but if a further differentiation seems necessary, de Vaucouleurs types are given (if available). Not counting individual galaxies in groups or clusters (though mentioned in the tables), a total of 500 objects are listed.

The data tables are followed by separate tables, which contain the textual descriptions. Around 600 descriptions are given based on the visual appearance of the object with different instruments: binocular (if possible), medium aperture telescope (6–10″), large aperture telescope (13–20″, sometimes even larger). To create a fairly homogenous set, the major part of observations is due to a small number of observers (unaccredited descriptions are from the authors, based on observations with various instruments, from 10×50 binoculars to a 36″ Dobsonian):

- Steve Gottlieb (SG): 8″, 13″, 17.5″, 18″, 20″
- Steve Coe (SC): 4.5″, 6″, 8″, 11″, 13.1″, 20″, 25″

To fill the remaining gaps, observations of a few other experienced observers were used: Jens Bohle (20″), Jeffrey Corder (12.5, 17.5″), Lynton Hemer (30″), Michael Kerr (8, 25″), Jeff Medkeff (10″), Tom Polakis (13, 25″ together with Larry Mitchell), Frank Richardsen (20″), Brian Skiff (6″), Auke Slotegraaf (15.5″), and Magda Streicher (8, 12″).

Chapter 7

Observing Programs

Every observing session requires a certain amount of preparation. The simplest question might be "What to observe?" The number of suitable targets for your telescope is large – perhaps too large. We have already described useful selection tools like star atlases, databases, software, and the Internet. How to reduce the "cluster" of galaxies in a practical manner? The selection can be based on individual criteria like brightness, size, position (visibility), distance, type, or certain structures. There are further possibilities to define the observing session: special catalogues (e.g., Messier, Herschel), or certain sky areas (e.g., conspicuous constellations). It is obvious that an individual object can occur in many different selections.

Whatever you choose, try to estimate the number of objects meeting your criteria. Be sure that the targets have a chance to be detected in your specific telescope. If your list tends to "explode," perhaps stronger restrictions are necessary. Large observing programs can be tedious and difficult to complete. Don't forget to answer questions like: When are the objects visible best? What is the adequate observing sequence? Is there any restriction due to the observing site? But most important is: Any observation should be satisfying – and produce no stress. So, not being a professional, don't be a slave of your program. Change it if necessary and feel free to interrupt it for sweeping through the sky with your binoculars!

Catalogue-Specific Observing

The beginner will often select objects just by their brightness. This can be done without much thinking: as the Messier catalogue is usually referred as the "biggest and brightest" object set in the sky. This is basic catalogue-specific observing. In this chapter we restrict our selections to the classic catalogues of nonstellar objects (M, NGC/IC) or galaxies (UGC). Further catalogues, e.g., specialized on certain types of galaxies, will be used later.

Messier Galaxies

Table 7.1 presents the data of the 40 galaxies contained in the extended Messier catalogue of 110 objects. Three can be seen with the naked eye, depending on the sky quality: M 31 (even under moderate conditions), M 33 (pretty dark sky, fst 6.3 mag or better), and M 81 (very dark sky, fst 7.5 mag). The faintest objects, according to their visual (integrated) magnitude V, are M 91 and M 98. All galaxies can be seen in a good pair of binoculars, and most require only a 7×50. Messier galaxies are ideal targets for the full range of telescope types and apertures [124].

Table 7.1. All 40 Messier galaxies

M	NGC	Con	R.A.	Decl	V	V'	$a \times b$	PA	Type	Remarks
M 31	NGC 224	And	00 42 44.3	+41 16 08	3.5	13.5	189.1 × 61.7	35	Sb	Andromeda Nebula
M 32	NGC 221	And	00 42 41.8	+40 51 57	8.1	12.5	8.5 × 6.5	179	E2	Arp 168
M 33	NGC 598	Tri	01 33 51.9	+30 39 29	5.5	14.0	68.7 × 41.6	23	Sc	Triangulum Nebula
M 49	NGC 4472	Vir	12 29 46.7	+08 00 00	8.3	13.2	10.2 × 8.3	155	E2	Arp 134
M 51	NGC 5194	CVn	13 29 52.6	+47 11 44	8.1	12.7	11.2 × 6.9	7	Sbc	Arp 85, Whirlpool, interacting with NGC 5195
M 58	NGC 4579	Vir	12 37 43.7	+11 49 06	9.6	13.1	6.0 × 4.8	95	SBb	Virgo Cluster
M 59	NGC 4621	Vir	12 42 02.2	+11 38 50	9.7	13.0	5.4 × 3.7	165	E5	Virgo Cluster
M 60	NGC 4649	Vir	12 43 39.8	+11 33 11	8.8	13.1	7.6 × 6.2	105	E2	Arp 116, pair w. NGC 4647, Virgo Cluster
M 61	NGC 4303	Vir	12 21 54.9	+04 28 22	9.3	13.1	6.5 × 5.9	162	SBbc	
M 63	NGC 5055	CVn	13 15 49.0	+42 01 59	8.5	13.2	12.6 × 7.2	105	Sbc	Sunflower galaxy
M 64	NGC 4826	Com	12 56 43.8	+21 40 59	8.5	12.7	10.0 × 5.4	115	Sab	Black Eye galaxy
M 65	NGC 3623	Leo	11 18 55.6	+13 05 27	9.2	12.7	9.8 × 2.9	174	Sa	Arp 317
M 66	NGC 3627	Leo	11 20 15.1	+12 59 24	8.9	12.7	9.1 × 4.1	173	Sb	Arp 16, Arp 317
M 74	NGC 628	Psc	01 36 41.7	+15 47 00	9.1	13.9	10.5 × 9.5	25	Sc	
M 77	NGC 1068	Cet	02 42 40.8	–00 00 46	8.9	12.8	7.1 × 6.0	70	Sb pec	Arp 37, Cetus A, brightest Seyfert galaxy
M 81	NGC 3031	UMa	09 55 33.5	+69 04 02	7.0	13.0	24.9 × 11.5	157	Sb	Bode's nebulae
M 82	NGC 3034	UMa	09 55 54.0	+69 40 59	8.6	12.7	11.2 × 4.3	65	Sd	Arp 337, Bode's nebulae, Ursa Major A
M 83	NGC 5236	Hya	13 37 00.2	–29 52 02	7.5	12.8	12.9 × 11.5	44	Sc	
M 84	NGC 4374	Vir	12 25 03.6	+12 53 13	9.2	13.2	6.5 × 5.6	135	E1	Virgo Cluster, Markarian's Chain
M 85	NGC 4382	Com	12 25 23.9	+18 11 27	9.1	13.0	7.1 × 5.5	5	S0-a	Virgo Cluster
M 86	NGC 4406	Vir	12 26 11.5	+12 56 47	8.9	13.3	8.9 × 5.8	130	E3	Virgo Cluster, Markarian's Chain
M 87	NGC 4486	Vir	12 30 49.4	+12 23 26	8.6	13.0	8.3 × 6.6	170	E pec	Arp 152, Virgo A, Virgo Cluster
M 88	NGC 4501	Com	12 31 59.0	+14 25 11	9.4	12.8	6.8 × 3.7	140	Sb	Virgo Cluster

M 89	NGC 4552	Vir	12 35 39.9	+12 33 22	9.9	12.7	3.5×3.5		E1	Virgo Cluster
M 90	NGC 4569	Vir	12 36 50.0	+13 09 50	9.4	13.3	9.5×4.4	23	SBab	Arp 76, Virgo Cluster
M 91	NGC 4548	Com	12 35 26.4	+14 29 47	10.1	13.3	5.2×4.2	150	SBb	
M 94	NGC 4736	CVn	12 50 53.1	+41 07 17	8.1	13.6	14.4×12.1	117	Sab	
M 95	NGC 3351	Leo	10 43 57.8	+11 42 12	9.8	13.6	7.4×5.0	13	SBb	
M 96	NGC 3368	Leo	10 46 45.8	+11 49 12	9.3	13.2	7.8×5.2	176	SBab	
M 98	NGC 4192	Com	12 13 47.8	+14 53 58	10.1	13.5	9.8×2.8	155	SBb	Virgo Cluster
M 99	NGC 4254	Com	12 18 49.3	+14 25 03	9.7	13.0	5.3×4.6	51	Sc	Pinwheel Galaxy, Virgo Cluster
M 100	NGC 4321	Com	12 22 54.9	+15 49 22	9.3	13.3	7.5×6.1	30	SBbc	Virgo Cluster
M 101	NGC 5457	UMa	14 03 12.4	+54 20 58	7.5	14.6	28.8×26.9	26	Sc	Arp 26
M 102	NGC 5866	Dra	15 06 29.4	+55 45 49	9.9	13.0	6.5×3.1	128	S0-a	
M 104	NGC 4594	Vir	12 39 59.3	−11 37 21	8.3	12.0	8.6×4.2	89	Sa	Sombrero Galaxy
M 105	NGC 3379	Leo	10 47 49.5	+12 34 52	9.5	13.1	5.3×4.8	71	E1	
M 106	NGC 4258	CVn	12 18 57.8	+47 18 25	8.3	13.5	18.6×7.2	150	SBbc	VV 448
M 108	NGC 3556	UMa	11 11 29.4	+55 40 22	9.9	13.0	8.6×2.4	80	Sc	
M 109	NGC 3992	UMa	11 57 35.4	+53 22 25	9.8	13.4	7.5×4.4	68	SBbc	
M 110	NGC 205	And	00 40 22.1	+41 41 07	7.9	13.8	19.5×11.5	170	E5	

Surface brightness V′ is often a more important criterion. According to this quantity, M 104 (Fig. 7.1)and M 32 are the brightest objects. At the faint end we meet M 33 and M 101. M 101 is too faint for the naked eye, and even in a binocular it is difficult. It is the third largest Messier galaxy (succeeding M 31 and M 33). The next smaller are M 81, M 110, and M 106. Many galaxies show peculiarities, like the famous M 51 system (Fig. 7.2), which can be detected with medium aperture (see Arp or VV designation). Visual descriptions with different apertures are given in Table 7.2.

Additional Notes

M 32. "The smallest giant." Though often classified as a "dwarf elliptical," its luminosity function is more in line with a giant elliptical. It has been suggested that it is actually the stripped off core of a much larger galaxy.

M 51. In 1845, Lord Rosse was the first to observe the spiral structure of this galaxy – leading to the term *spiral nebulae.* This is a classic example of an interacting galaxy, as the smaller galaxy (NGC 5195) has distorted and pulled out a long tidal plume (Fig. 7.2). This is the archetype of a group of tidally distorted galaxies known as "Whirlpool Galaxies" that includes systems such as NGC 1531/2 (Fig. 7.3), NGC 5216/8 and NGC 7752/3 .

M 60. This galaxy and NGC 4647 in Virgo are a classic example of an "overlapping," non-interacting pair (see Table 9.2). They and others like NGC 2207 + IC 2163 in Canis

Fig. 7.1. M 104 in Virgo, a galaxy with pretty high surface brightness

Fig. 7.2. One of the most prominent Messier galaxies: M 51 in Canes Venatici, interacting with NGC 5195

Major and the Siamese Twins (NGC 4567/8 in Virgo) lie in the same line of sight and are not true physical systems.

M 74. A beautiful "grand design" spiral, with numerous large HII regions and OB associations strewn along the spiral arms.

M 77. A Seyfert Type II galaxy (AGN), the surface brightness of the core region of this galaxy is very high (Fig. 1.27). Lord Rosse once called this the "Blue Galaxy," due to the bluish tint he observed with the 72 in. reflector.

M 81 and *M 82.* One of the brightest galaxy pairs in the heavens. M 81 has been spotted by the naked eye by advanced observers under extremely dark skies (Fig. 5.5). M 81 is a "grand design" spiral, while the much more distorted M 82 is undergoing a massive starburst (Fig. 1.21).

M 83. A grand design barred spiral (Fig. 3.3), its spiral structure was resolved by both Lord Rosse and William Lassell in the 1840s. Under very dark skies, the spiral structure may be observed with a 10–12 in. scope. Over the past 50 years, this galaxy has produced numerous bright supernovae.

M 87. The most massive galaxy of the Messier list, this giant elliptical lies near the center of the Virgo Cluster. It hosts a massive black hole in the core, which produces a jet that may be seen at high power in scopes as small as 16 in. Though difficult to observe, this structure and the brighter globular clusters can be resolved by a modest telescope equipped with a CCD.

Table 7.2. Visual descriptions of the Messier galaxies using different apertures

M	Binocular	Medium aperture	Large aperture
31	Bright center with oval halo (7 × 50)	Bright center with large, oval halo; two dark lanes at 100× (8″)	Bright nucleus and structured disk with two dark lanes and knots, e.g., NGC 206 in SW (14″)
32	Easily visible, stellar (10 × 50)	Very bright, compact nucleus, diffuse edge with no structures (8″, 200×)	Very bright, moderately large, elongated 4:3 NNW–SSE, about 4′ × 3′, increases to small very bright core which is almost stellar (SG 13″)
33	Oval disk with no structure, no nucleus (16 × 70)	Elongated, low surface brightness object, small nucleus, no structure, except NGC 604, which is difficult (8″)	Very large object, spiral arms can be traced by their brightest knots; NGC 604 is easy (14″)
49	Faint, round (7 × 50)	Fairly bright, round, stellar core with smooth halo (8″)	Very bright, fairly large, sharp concentration to a compact very bright nucleus, large halo slightly elongated ~N–S fades at the edges. A 12 mag star is superimposed at the E edge 0.8′ from center (SG 17.5″)
51	Clearly visible as small cloud (16 × 70)	Conspicuous object at 100×; bright nucleus; at 200× spiral arms with structures and connection to companion NGC 5195 (bright nucleus with halo) (8″)	Bright nucleus with two spiral arms and knots; companion irregular with bright nucleus and extension (14″)
58	Faint, oval patch (16 × 70)	Pretty bright, stellar core with oval halo (8″)	Bright, moderately large, slightly elongated 4:3 WSW–ENE, small very bright core, stellar nucleus (SG 17.5″)
59	Faint, oval patch (16 × 70)	Bright center, elongated. In triangle with two stars (8″)	Very bright, moderately large, oval NNW–SSE, 3′ × 2′, small very bright core, stellar nucleus. A 15 mag star is at the SW edge and a brighter 13 mag star is off the N end 1.9′ from center (SG 17.5″)
60	Pretty easy, round (16 × 70)	Bright nucleus; diffuse halo with no structure; companion NGC 4647 not visible (8″, 100 ×)	Round, structureless disk with bright center; companion clearly visible (14″)
61	Faint, round, diffuse patch (16 × 70)	Bright with stellar core, oval halo with weak structures (8″)	Two or three arms visible, interesting structure (SG 17.5″)
63	Diffuse spot (16 × 70)	Conspicuous, elongated object with bright nucleus at 100×; at 150× dark structure barely visible (8″)	Very bright, large, elongated 2:1 WNW–ESE, 6′ × 3′. There is a faint outer extension to the WNW (outer spiral arms?) which reaches extremely close to 8.7 mag star (SG 17.5″)
64	Visible with averted vision (10 × 50)	Bright, large, dark lane near the center, oval halo (8″, 100×)	Very bright, small nucleus, "black eye" sharply defined, two spiral arms to the edge with dark lane south (18″, 360×)

65	Faint oval spot (10×50)	Much elongated with bright nucleus; weak structures at 150× (8″)	Very bright, very large, very elongated N–S, 7.5′×2.0′, bright core, stellar nucleus. A 12 mag star is W of the S end 2.1′ from the center (SG 17.5″)
66	Faint diffuse patch (10×50)	Bright center with elongated halo; at 150× knots visible (8″)	Very bright, large, elongated N–S, 5′×3′, bright elongated core. Two spiral arms are visible although the western arm is more prominent (SG 17.5″)
74	Fairly faint, diffuse patch (16×70)	Very small bright core surrounded by a large faint halo (SG 8″)	Bright, large, round, very bright core. A spiral arm is attached at the E side of core winding toward the W along the S side. A dark gap is visible between the arm and the main central portion. Several stars are superimposed in the halo (SG 17.5″)
77	Faint, stellar (10×50)	Intense core, faint halo (SG 8″)	Very bright, moderately large, sharp concentration with an unusually bright core, almost stellar nucleus, diffuse slightly elongated halo. Appears mottled at high power and a hint of inner arm structure (SG 17.5″)
81	Easy, oval, diffuse (10×50)	Very bright, bright core, large oval halo, elongated NW–SE, two faint stars involved (SG 8″)	Very bright, very large, elongated 2:1 NNW–SSE, about 16′×8′, large oval bright middle, bright core, nearly stellar nucleus. Two stars of 11.5 mag and 11.9 mag are superimposed in the halo at the S edge of the core. An easily visible spiral arm is attached near these two stars at the S end of the core. This arm curves due N along the E side and is well separated from the main body. A second arm is suspected as a short extension curving around the NNW end toward a 12 mag star at the WNW edge of the halo (SG 17.5″)
82	Faint, elongated (16×70)	Bright, spindle, mottled. A dark wedge cuts into the galaxy near the center from the S side (SG 8″)	Very bright, large, edge-on 4:1 WSW-ENE, 10′×2.5′, large bright irregular core. Very mottled with an unusually high surface brightness. Unique appearance with several dark cuts oblique to the major axis including a prominent wedge or cut nearly through the center. A 10 mag star is just S of the SW end 5.8′ from the center 13; two obvious dark lanes (SG 17.5″)
83	Very faint, large, diffuse (16×70)	Bright, large, diffuse patch, mottled, round core (8″)	Spiral arms obvious, core blazing with 3″–4″ in size, many bright areas in arms (SC 20″)

(*Continued*)

Table 7.2. Visual descriptions of the Messier galaxies using different apertures—Cont'd

M	Binocular	Medium aperture	Large aperture
84	Faint, round patch (10×50)	Pretty bright, slightly elongated, stellar nucleus (8″)	Very bright, moderately large, almost round, very bright core, very small bright nucleus, halo gradually fades into background sky so there is no sharp edge. Nearly an identical twin of M 86 17′ ENE but rounder (SG 17.5″)
85	Easy, star south (10×50)	Bright, round, without structure, star superimposed (8″)	Very bright, moderately large, small very bright core. A 13 mag star is superimposed near the NNE edge and a 10 mag star is off the SE side 2.7′ from center (SG 17.5″)
86	Faint, round patch (10×50)	Smooth round patch, pretty bright, similar to M 84 (8″)	Very bright, fairly large, slightly elongated 4:3 NW–SE, 4′×3′, intense core, substellar nucleus, large diffuse halo (SG 17.5″)
87	Easy, round (10×50)	Very bright, stellar core with diffuse round halo (8″)	Very bright, fairly large, gradually increases to a very bright core, no sharp nucleus (SG 17.5″). The famous "jet" is visible (Frank Richardsen 20″, 850×)
88	Fairly faint, round (10×50)	Bright, elongated patch, stellar nucleus (8″)	Very bright, very large, elongated 5:2 NW–SE, brighter core, intense very small or stellar nucleus. A faint double star is embedded at the SE end (SG 17.5″)
89	Faint, round (16×70)	High surface brightness, small, round patch (8″)	Very bright, irregularly round, fairly small but high surface brightness with an intense, very small bright core and substellar nucleus (SG 17.5″)
90	Fairly faint patch (16×70)	Bright, elongated, star superimposed near stellar (8″)	Bright, large, very elongated 3:1 SSW–NNE, sharp concentration as suddenly increases to a bright stellar nucleus (possibly a superimposed star), fairly even surface brightness to halo (SG 17.5″)
91	Very faint spot (16×70)	Bright, slightly elongated, structured halo (8″)	Bright, moderately large, elongated 3:2 SW–NE, 3′×2′, gradually increases to a bright core and a very small nucleus (SG 17.5″)
94	Faint, stellar (10×50)	Bright, structured core, much elongated patchy halo (8″)	Very bright, pretty compact, slightly elongated, stars in halo (14″)
95	Fairly faint (16×70)	Bright, fairly large, round (SG 8″)	Very bright, very bright core. The outer halo is 4.5′×3.0′ oriented SSW–NNE. A bar is highly suspected extending WNW–ESE of the central core with inner ring structure suspected extending from this bar (SG 17.5″)

96	Faint, diffuse (16 × 70)	Bright, fairly large, slightly elongated (SG 8″)	Very bright, fairly large, elongated NW–SE, 5′ × 3.5′, small bright core, stellar nucleus (SG 17.5″)
98	Very faint (10 × 50)	Very bright, large elongated, gradually brighter middle (8″)	Bright, very large, very elongated 4:1 NNW–SSE, 6′ × 1.5′, small bright core, stellar nucleus. A faint knot is highly suspected near the S tip (SG 17.5″)
99	Small round patch (10 × 50)	Bright, large round core, mottled halo, sprial arms extends to E, star at the SE edge (8″)	Very bright, large, bright core, stellar nucleus. There is an obvious spiral arm attached at the SE side of the core and winding along the S side toward the W. There is a dark gap between the spiral arm and the core along the S and W side. A second shorter, diffuse arm is visible on the N side (SG 17.5″)
100	Fairly faint, diffuse (16 × 70)	Bright, large with bright core and faint halo (8″)	Bright, very large, almost round, well-defined bright core surrounded by a large, fainter halo (SG 17.5″)
101	Vague, round patch without any structure (16 × 70)	Difficult if sky is not dark; LPR filter helps to enhance the low contrast; large round with diffuse edge and weak nucleus (8″, 80×)	Large round area with spiral arms and knots, nucleus weakly concentrated (14″)
102	Vague, slightly elongated patch, no structure (16 × 70)	Fairly bright, lens-shaped in NW–SE, center round and brightened (8″)	Bright, lens-shaped galaxy, with bright elliptical core 1′ × 0.5′, covered by a small dark lane; star 15 mag north of center (20″)
104	Easy, slightly elongated patch (10 × 50)	Bright round nucleus with oval halo and thin dark lane south, small brightening further south (8″)	Oval spindle with bright nucleus, cut by an slightly curved absorption band; northern part much brighter (14″)
105	Fairly faint spot (10 × 50)	Fairly bright, round (SG 8″)	Bright, very small bright core, slightly elongated. First of three bright galaxies in the field with NGC 3384 7.3′ NW and NGC 3389 9.7′ ESE (SG 17.5″)
106	Faint, elongated (10 × 50)	Bright, very large, elongated, bright core (SG 8″)	Very bright, very large, very elongated 3:1 NNW-SSE, 14′ × 4′, large bright core concentrated to a very small brighter central region. A thin bright spiral arm attached at the core extends toward the NNW on the following side of the galaxy. There is a sharp edge along the W side of this arm (SG 17.5″)

(*Continued*)

Table 7.2. Visual descriptions of the Messier galaxies using different apertures—Cont'd

M	Binocular	Medium aperture	Large aperture
108	Fairly faint, elongated (16 × 70)	Bright, large gradually brighter core, edge-on, star superimposed W of center (8″)	Very bright, very large, edge-on 4:1 WSW–ENE, 8′ × 2′. A 12 mag star is superimposed just W of center appearing similar to a bright stellar nucleus. Two fainter stars are also superimposed E of the core. A bright knot is visible W of the core (1.3′ W of the star) and the region near the core appears dusty. A 12 mag star is just S of the W end 4.9′ from the center (SG 17.5″)
109	Very faint, oval (16 × 70)	Faint, diffuse halo, small core, slightly elongated (8″)	Bright, large, elongated 5:3 SW–NE, at least 6′ × 3.5′, broadly concentrated halo, large faint halo. A 13 mag star is superimposed on the halo 50″ NNW of center. A 13 mag star is at the NE edge of the halo 3.4′ from center (SG 17.5″)
110	Oval disk, much fainter than M 32, no structures (16 × 70)	Oval diffuse disk with small, round nucleus, no structures (8″)	Large diffuse object, with moderately low surface brightness, not much concentrated (14″)

Fig. 7.3. NGC 1532 in Eridanus, in pair with NGC 1531

M 101. A giant Sc type galaxy (Fig. 4.4), this system is far larger and more massive than our Milk Way. Many of the massive HII regions than help define the arms are visible with moderate (12 in.) aperture.

Bright NGC/IC Galaxies

If you leave out the "Herschel 400" [125], it's likely that your "next" catalogue will be the *New General Catalogue* (NGC) and its appendix, the *Index Catalogue* (IC). There will be around 1,000 NGC/IC galaxies visible in a 10″ telescope, and perhaps 6-times more in a 16″. Who wants to see them all, should start soon. A number of top observers have seen several thousand of these objects, and near the top is contributor Steve Gottlieb, credited for more than 6,800 observed NGC/IC objects (most of them are galaxies)! Half as much, but still a lot, has been visited by the German amateur Klaus Wenzel.

Due to the large number of NGC/IC galaxies, we can only present a small sample, selected by visual magnitude (V) from the *Revised New General and Index Catalogue* [74]. Table 7.3 lists northern galaxies ($\delta \geq -30°$), Table 7.4 southern galaxies ($\delta < -30°$). It is interesting that there are many bright non-Messier galaxies, e.g., NGC 3628 in Leo (Fig. 7.4) . Too far south for Messier is bright NGC 300 in Sculptor (Fig. 7.5) . Note that the *Revised Shapley-Ames Catalog of Bright Galaxies* [81], containing all galaxies brighter than B_T = 13.2 mag, is a valuable source for galaxy observations with an 8″ telescope. Visual descriptions of bright NGC/I C galaxies are given in Table 7.5.

Note that the Large Magellanic Cloud (LMC, Nubecula major) is not included as it bears no NGC number. The LMC is the largest and brightest galaxy on the sky (except the Milky Way) and is easily visible to the naked eye, but unfortunately not for northern observers. The LMC and SMC are "single" objects only in the smallest telescopes. In a 4 in., both become "multiple" – a large reservoir of catalogued structures (many with NGC/IC designations). That's why we will not describe the clouds here. Who wants to know all about them, should first consult the work of Jenni Kay [126] or Vol. 7 ("The Southern Sky") of the Webb Society Handbook.

Table 7.3. Northern NGC/IC galaxies brighter than $V = 10$ mag (not listed in the Messier catalogue)

NGC/IC	Con	R.A.	Decl	V	V'	$a \times b$	PA	Type	Remarks
NGC 185	Cas	00 38 57.6	+48 20 14	9.3	13.7	8.0×7.0	35	E3	M 31 group
NGC 247	Cet	00 47 08.3	−20 45 36	8.9	13.8	19.2×5.5	172	SBcd	Sculptor group
NGC 253	Scl	00 47 33.1	−25 17 15	7.3	12.9	29.0×6.8	52	SBc	Silver Dollar Galaxy, Sculptor group
IC 1613	Cet	01 04 54.2	+02 08 02	9.3	15.1	16.6×14.9	50	IBm	Local Group member
NGC 613	Scl	01 34 18.4	−29 25 07	9.9	13.2	5.5×4.2	120	SBbc	VV 824
NGC 1023	Per	02 40 24.1	+39 03 48	9.5	12.7	7.4×2.5	87	E/SB0	
NGC 1232	Eri	03 09 45.3	−20 34 45	9.8	13.9	7.4×6.5	108	SBc	Arp 41
NGC 1395	Eri	03 38 29.6	−23 01 38	9.8	13.3	5.0×4.5	120	E2	
NGC 1398	For	03 38 52.0	−26 20 14	9.8	13.6	7.2×5.2	100	SBab	
NGC 1407	Eri	03 40 11.8	−18 34 49	9.7	13.0	4.6×4.3	35	E	
IC 342	Cam	03 46 48.4	+68 05 44	8.4	14.9	21.4×20.9	168	SBc	Beyond the Local Group
NGC 2683	Lyn	08 52 41.3	+33 25 12	9.7	12.8	9.3×2.1	44	Sb	
NGC 2841	UMa	09 22 02.3	+50 58 35	9.3	12.8	8.1×3.5	147	Sb	
NGC 2903	Leo	09 32 09.7	+21 29 57	8.8	13.3	12.6×6.0	17	SBbc	
NGC 3115	Sex	10 05 14.1	−07 43 05	9.1	12.3	7.2×2.4	40	E-S0	Spindle Galaxy
NGC 3184	UMa	10 18 17.0	+41 25 24	9.6	13.7	7.4×6.9	135	SBc	
NGC 3344	LMi	10 43 30.9	+24 55 22	9.7	13.7	7.1×6.5	18	SBbc	
NGC 3384	Leo	10 48 16.7	+12 37 43	9.9	12.9	5.4×2.7	53	E/SB0	NGC 3371
NGC 3521	Leo	11 05 48.8	−00 02 13	9.2	13.5	11.2×5.4	163	SBbc	
NGC 3585	Hya	11 13 17.3	−26 45 18	9.9	12.6	4.6×2.5	107	E6	
NGC 3607	Leo	11 16 54.5	+18 03 08	9.9	13.1	4.6×4.0	120	E-S0	
NGC 3628	Leo	11 20 16.7	+13 35 24	9.6	13.5	13.1×3.1	104	Sb	In Leo Triplet
NGC 3953	UMa	11 53 48.4	+52 19 30	9.8	13.1	6.9×3.6	13	SBbc	
NGC 4125	Dra	12 08 05.5	+65 10 28	9.6	12.8	5.8×3.2	81	E6 pec	
NGC 4214	CVn	12 15 38.8	+36 19 39	9.6	13.8	8.0×6.6	144	IBm	NGC 4228
NGC 4449	CVn	12 28 11.3	+44 05 42	9.4	12.8	6.2×4.4	45	IBm	Beyond the Local Group

NGC 4490	CVn	12 30 36.1	+41 38 34	9.5	12.6	6.4×3.2	125	SBcd	Pair w. NGC 4485 (Arp 269)
NGC 4494	Com	12 31 24.1	+25 46 31	9.7	12.8	4.8×3.5	171	E1	
NGC 4526	Vir	12 34 02.8	+07 41 56	9.6	12.6	7.0×2.5	113	SB0	NGC 4560, Virgo Cluster
NGC 4535	Vir	12 34 20.2	+08 11 51	9.8	13.5	7.1×5.0	0	SBc	Virgo Cluster
NGC 4559	Com	12 35 57.8	+27 57 35	9.6	13.6	10.7×4.4	150	SBc	
NGC 4565	Com	12 36 20.5	+25 59 16	9.5	13.2	15.8×2.1	136	Sb	
NGC 4631	CVn	12 42 07.6	+32 32 30	9.0	12.9	15.2×2.8	86	SBcd	Arp 281
NGC 4636	Vir	12 42 49.7	+02 41 14	9.4	13.1	5.9×4.6	150	E	
NGC 4697	Vir	12 48 35.8	–05 48 00	9.2	13.1	7.2×4.7	70	E6	
NGC 4699	Vir	12 49 02.2	–08 39 50	9.6	12.0	3.8×2.8	45	SBb	
NGC 4725	Com	12 50 26.5	+25 30 00	9.3	13.9	10.7×7.6	35	SBab	
NGC 4753	Vir	12 52 22.1	–01 12 00	9.9	12.8	6.0×2.8	80	S0	
NGC 5005	CVn	13 10 56.1	+37 03 31	9.8	12.7	5.8×2.9	65	SBbc	
NGC 5068	Vir	13 18 54.5	–21 02 17	9.8	13.8	7.3×6.4	110	SBc	
NGC 5247	Vir	13 38 02.9	–17 53 05	9.9	13.3	5.4×4.9	20	SBbc	
NGC 5866	Dra	15 06 29.4	+55 45 49	9.9	13.0	6.5×3.1	128	S0-a	M 102
NGC 6946	Cyg	20 34 52.1	+60 09 12	9.0	14.0	11.5×9.8	57	SBc	
NGC 7331	Peg	22 37 05.1	+34 25 13	9.5	13.4	10.2×4.2	171	Sbc	

Table 7.4. Southern NGC galaxies brighter than $V = 10$ mag

NGC	Con	R.A.	Decl	V	V'	$a \times b$	PA	Type	Remarks
NGC 55	Scl	00 15 08.0	–39 13 10	7.8	13.3	31.2×5.9	108	SBm	Sculptor group
NGC 300	Scl	00 54 53.3	–37 41 03	8.1	13.9	19.0×12.9	111	Scd	Sculptor group
NGC 1097	For	02 46 19.5	–30 16 32	9.5	13.8	9.4×6.6	130	SBb	Arp 77
NGC 1291	Eri	03 17 18.3	–41 06 26	8.5	13.4	11.0×9.5	72	SB0-a	NGC 1269
NGC 1313	Ret	03 18 16.0	–66 29 43	9.1	13.5	9.2×7.2	38	SBcd	VV 436
NGC 1316	For	03 22 41.4	–37 12 28	8.4	13.0	11.0×7.2	50	SB0	Arp 154, Fornax A
NGC 1365	For	03 33 36.7	–36 08 27	9.5	13.9	11.0×6.2	32	SBb	VV 825
NGC 1380	For	03 36 27.5	–34 58 31	9.9	12.2	4.0×2.4	7	SB0	
NGC 1433	Hor	03 42 01.2	–47 13 19	9.8	13.6	6.5×5.9	99	SBa	
NGC 1532	Eri	04 12 03.8	–32 52 23	9.8	13.6	11.6×3.4	33	SBb	Pair w. NGC 1531
NGC 1549	Dor	04 15 45.0	–55 35 29	9.6	12.9	4.9×4.1	135	E0	Pair w. NGC 1553
NGC 1553	Dor	04 16 10.6	–55 46 46	9.0	11.6	4.5×2.8	150	S0	Pair w. NGC 1549
NGC 1566	Dor	04 20 00.5	–54 56 14	9.4	13.6	8.2×6.5	60	SBbc	Seyfert galaxy
NGC 1672	Dor	04 45 42.8	–59 14 52	9.7	13.5	6.7×5.6	170	SBb	VV 826
NGC 1808	Col	05 07 42.5	–37 30 48	9.9	13.3	6.5×3.9	133	SBa	
NGC 2997	Ant	09 45 38.6	–31 11 26	9.4	13.7	8.9×6.8	110	SBc	
NGC 4945	Cen	13 05 26.1	–49 27 46	8.6	13.2	19.8×4.0	43	SBc	
NGC 5102	Cen	13 21 57.0	–36 37 54	9.5	13.0	8.6×2.7	48	E-S0	
NGC 5128	Cen	13 25 29.0	–43 00 58	6.6	13.3	25.7×20.0	35	S0	Arp 153, Centaurus A
NGC 7793	Scl	23 57 49.2	–32 35 30	9.0	13.3	9.3×6.3	98	Scd	Sculptor group

Fig. 7.4. Near to M 65 and M 66, but not seen by Messier: NGC 3628 in Leo

Fig. 7.5. A bright southern galaxy: NGC 300, a member of the Sculptor group

Table 7.5. Visual description of NGC/IC galaxies (from Table 7.3 and Table 7.4) using different apertures; sorted by NGC/IC number (Local Group members are described in Table 8.2, IC 342 in Table 8.4)

NGC	Small/Medium aperture	Large aperture
55	Pretty bright, large, very much elongated in PA 120°. Somewhat brighter middle, pretty low surface brightness (SC 8″)	Viewed at nearly 60° elevation at 212×, this huge galaxy was an amazing sight and overfilled the 23′ field (at least 25′ in length). Near the core were two small, prominent HII knots. A couple more low surface brightness knots were visible further out on the mottled extensions. The appearance was asymmetric with the brighter WNW section bulging slightly (SG 20″)
247	Large, faint, diffuse patch (16×70). Very large, elongated ~N–S, bright core. A 10 mag star is at the S tip. The S extension appears brighter (SG 8″)	Bright, very large, bright core, elongated 7:2 N–S, 14′×4′. The southern extension is brighter and a 9 mag star is superimposed at the S end about 6′ from the core (SG 17.5″)
253	Easy spindle (16×70). Large elongated, bright object, compact center with bar, arms, and dust structures slightly visible (10″, 100×)	Many details; bright spiral arm NE–SE; dark knots near center; dark lane SE of center, cutting the inner and outer edge (18″, 150×)
300	Fairly bright, fairly large, oval 3:2 WNW–ESE, very diffuse, bright stellar nucleus. There is a hint of structure but has a low surface brightness (SG 13″)	Faint, large, several star superimposed, somewhat brighter in the middle at 100×. This is a low surface brightness object. Imagine only 10° above the horizon (SG 17.5″). Large, nearly face-on spiral with two main arms and hints of others. Numerous HII regions are visible, lending the appearance of a smaller, fainter version of M 33 (RJ 24″)
613	Faint, moderately large, diffuse, small bright core. A 9 mag star is 2.5′ NE (SG 8″)	Fairly bright, very elongated 3:1 WNW–ESE, 4.0′×1.3′, prominent elongated core, almost stellar nucleus with direct vision, broader halo with averted vision. SE of the core there appears to be a very faint extension or large knot (SG 17.5″)
1023	Fairly bright, bulging bright core, lens-shape (SG 8″)	Bright, large, very elongated 7:2 ~E–W, very bright core, almost stellar nucleus. Large fainter halo increases size to 7′×2′. Two 15 mag stars are superimposed on the W and E ends (SG 17.5″)
1097	Bright, elongated NW–SE, bright core (SG 8″)	Very bright, very large, very elongated NW–SE, very bright core. A companion galaxy NGC 1097A is attached at the NW end (SG 17.5″)
1232	Faint, pretty small, much brightermiddle, round. Not much in 6″, never saw any arm detail. Averted vision makes it larger (SC 6″)	Bright, large, slightly elongated, bright core, very large faint halo (SG 17.5″)
1291	Visible in 15×70.	Very bright, moderately large, large very bright core. A 12 mag star is just off the N end 1.7′ from the center (SG 17.5″)
1313	Fairly bright, oval haze, stars superimposed (8″)	Pretty bright, large, slightly elongated, mottled, patchy structure (14″)

1316	Bright, round, slightly elongated, small bright core. Forms a pair with NGC 1317 7′ N (SG 8″)	Very bright, moderately large, elongated 3:2 SW–NE, about 2.5′ × 1.5′. Dominated by an intense 40″ × 30″ core which brightens to a nonstellar nucleus (SG 17.5″)
1365	Fairly bright, fairly large, bright core, diffuse halo, broad concentration (SG 8″)	Bright, elongated core, large, 3′ diameter, very diffuse outer halo (SG 13″). Very bright, pretty large, spiral structure obvious; dark lane in central core there about 50% of the time; core is elongated 1.8′ × 1′; dark lane cuts core in ⅓ and pieces; several bright knots and some mottling in arms which extend out from central bar (SC 25″)
1380	Fairly bright, moderately large, elongated, bright core (SG 8″)	Very bright, elongated 2:1 N–S, bright core, faint elongated halo. A very faint 14 mag star is SW of the core 1.2′ from the center. Member of Fornax cluster (SG 13″)
1395	Fairly bright, small, round, small bright core (SG 8″)	Bright, fairly small, oval 4:3 ~E–W, very bright core, fainter halo. Two faint 14 mag stars lie on the W and N edges 1.0′ from center (SG 13″)
1398	Fairly bright, moderately large, round, bright core (SG 13″)	Very bright, moderately large, elongated 2:1 N–S, 2.2′ × 1.1′, well concentrated with a very bright 30″ rounder core and a stellar nucleus (SG 17.5″).
1407	Bright, small, round, small bright core (SG 8″)	Bright, fairly small, bright core, stellar nucleus. Forms a wide pair with NGC 1400 11.6′ SW (SG 13″)
1433	Large, faint, small bright nucleus (8″)	Bar clearly visible while nuclear region appears bright and nonstellar (14″)
1532	Elongated, bright oval core (8″)	Pretty bright, pretty large, very elongated with a bright nucleus at 135×. This edge-on and the round NGC 1531 make an interesting pair (SG 17.5″)
1549 1553	Two beautiful smudges of nebulous halos, clearly seen as fuzzy patches while sweeping. NGC 1549 is fainter, just outside a half-degree triangle; round glow; seems to be one or more small stars involved close to the galaxy. NGC 1553 small, round, pretty bright glow, lies on one leg of a half-degree triangle (Auke Slotegraaf 11 × 80)	NGC 1549 round, small and fairly bright with a very bright off-centre nucleus, and faint stars to the west involved close by. The field of view is been scattered with numerous faint stars which made it exceptional. An 8 mag star is been situated to the immediate south on its way to NGC 1553, situated only 11′ to the south. Together they are visible in 42.2′ field of view. NGC 1553 is a small clearly oblong-elongated (N–S) relatively bright galaxy, gradually brightening to a moderate yet small nucleus. The northern section of the galaxy appears to fade into a haze. Three field stars forming a triangle with two close to the centre of the galaxy situated to the west (Magda Streicher 12″). NGC 1549 looks like a globular cluster, round and diffuse. The galaxies central condensation is slightly off-centre. It lies near some bright stars. NGC 1553 lies in the same low-power field. NGC 1553 appears bright, large, and clearly elongated. It's northern tip ends in a star, while the southern tip appears to fade more gradually. It has a clear, small centre (Auke Slotegraaf 15.5″, 220×)

(*Continued*)

Table 7.5. Visual description of NGC/IC galaxies (from Table 7.3 and Table 7.4) using different apertures; sorted by NGC/IC number (Local Group members are described in Table 8.2, IC 342 in Table 8.4)—Cont'd

NGC	Small/Medium aperture	Large aperture
1566	Seen with some attention, to the south-east of a small star. Like a tiny globular cluster; east-southeast of a star lies this obvious small, globular-cluster like galaxy (Auke Slotegraaf 11 × 80)	Large, slightly elongated face-on spiral (NE-SW) galaxy with a misty appearance and fleecy edges. Rising slowly to a bright nucleus. The centre of the galaxy is noticeably wide so much so that it covers one third of the entire galaxy. A slightly yellow 8 mag star is visible about 3′ to the north-west in a bare star-field. A 13 mag star seen easily imbedded in the eastern hazy edge of the galaxy especially with averted vision. The NE-SW elongated sides of the galaxy looks somewhat more hazy which indicated the spiral structure (Magda Streicher 12″). The galaxy shows as a bright, diffuse glow, circular, with a definite nucleus. Using averted vision, the galaxy looses its circular shape, and takes on the appearance of an extended oval. There is a small star near the southern tip of the galaxy. The western edge seems clearly marked with a straight-edge (Auke Slotegraaf 15.5″, 220×)
1672	Fairly bright, small, round, brighter middle, in rich star field (8″)	This striking spiral galaxy appeared fairly bright and large, ~4′ diameter, sharply concentrated with a very bright core. Clearly emerging from the east side of the oval core or bar was a spiral arm which curled north and wrapped around two stars to the NW of the core. The extension on the west side was just a very faint, diffuse haze on the SW side without arm structure (SG 18″)
1808	Fairly bright, elongated NW–SE, moderately large, bright core (SG 8″)	Bright, fairly large, small elongated core, long thin arms 4:1 NW–SE. A 14 mag star is off the NW end. This is a very pleasing galaxy (SG 17.5″)
2683	Elongated, brighter middle (16 × 70). Small streak, weakly defined center without nucleus (8″, 100×)	Very long, slightly asymmetric nucleus with oval halo; weak dust lane (14″, 266×)
2841	Bright, small core, smooth oval halo, dark streak estimated (8″)	Bright, large, very small very bright nucleus, elongated 2:1 NW–SE, 6′ × 3′. There is a sharp light cut-off on the E side due to dust (SG 17.5″)
2903	Visible as faint patch (10 × 50). Bright, large, elongated, bright mottled core (SG 8″)	Very bright, very large, elongated 5:2 SSW–NNE, 10′ × 4′. A very faint knot is involved on the NNE side 1.2′ from center = NGC 2905. An extremely faint knot is also symmetrically placed opposed the core on the SW end 1.2′ from center. Dusty, mottled appearance with knots and arcs easy with averted vision (SG 17.5″)
2997	Bright, large, oval, low surface brightness with compact core (8″)	Fairly bright, very large, elongated 3:2 ~E–W, 4.5′ × 3.0′, sharply concentrated with a bright core, no nucleus. A 13 mag star is at the SSW edge of the halo 2.0′ from center (SG 13″)
3115	Faint oval spot (16 × 70). Very bright, high surface brightness, very bright core (SG 8″)	Very bright, fairly large, edge-on spindle 3:1 SW–NE, 5.5′ × 1.8′. Unusually high surface brightness, bright core, stellar nucleus (SG 17.5″)

3184	Faint, diffuse, weakly concentrated, star N within halo (8″)	Fairly bright, large, slightly elongated ~N–S, large 4′ halo has a fairly low surface brightness, very weak concentration, small brighter elongated core. A 11.5 mag star is at the N edge of the halo 1.8′ from the center. There is an impression of spiral structure but it is not distinct (SG 17.5″)
3344	Faint, large, low surface brightness. Two 10 mag stars are at the E edge (SG 8″)	Fairly bright, large, about 4′ × 3′ extended ~E–W. Unusual appearance as two bright stars are involved on the E side. Sharp concentration with a faint outer halo and a well-defined much brighter core. A 10.5 mag star is on the E side 52″ from the center and a 10 mag star is at the E edge of the halo 1.6′ from the center. Also a 13.5 mag star is superimposed about 30″ SE of the core (SG 17.5″)
3384	Faint patch near M 105 (16 × 70). Fairly bright, round, moderately large (SG 8″)	Bright, bright stellar nucleus, elongated 5:2 SW–NE (SG 13″)
3521	Pretty bright, elongated, stellar nucleus. (8″)	Very bright, very large, elongated 5′ × 2′ NNW–SSE. This is an impressive galaxy! Well-defined small bright oval core NNW–SSE, stellar nucleus. Appears mottled near the core and on the W side. Along the W side is a dust lane evident as a sharp light cut-off. The W side is somewhat fainter due to dust but extends beyond the dust lane (SG 17.5″)
3585	Bright, oval spot (8″)	Very bright, fairly small, elongated 2:1 WNW–ESE, very high surface brightness, very bright core, stellar nucleus (SG 17.5″)
3607	Bright, elongated, star superimposed (8″)	Bright, slightly elongated, bright core, stellar nucleus (SG 13″)
3628	Very faint (16 × 70). Bright, elongated halo, dark structure barely visible (8″)	Bright, unusually large edge-on WNW–ESE, 11′ × 2.5′. A broad irregular dust lane is prominent bisecting the galaxy along the entire length. Appears brighter to the N of the dark lane and fainter on the S side (SG 17.5″)
3953	Visible in 16 × 80 finder	Very bright, very large, elongated ~N–S, 5′ × 2′, very bright core, stellar nucleus. A 13 mag star is at the W edge 0.9′ from the center and a brighter 11 mag star is off the NE side 2.7′ from center (SG 17.5″)
4125	Pretty bright, elongated, stellar nucleus, star near (8″)	Bright, moderately large, very elongated almost 4:1 E–W, 2.5′ × 0.7′. A very bright elongated core and nearly stellar nucleus dominates the galaxy with much fainter extensions but overall the surface brightness is high. A 10 mag star is 2.4′ ESE of center. Forms a pair with NGC 4121 3.6′ SW (SG 17.5″)
4214	Pretty bright, pretty large, elongated 1.5′ × 1′ in PA 120°. Has a brighter middle (SC 6″)	Bright, large, slightly elongated NW–SE, bright core. There is a strong impression of curvature at the ends of the major axis (SG 13″)

(*Continued*)

Table 7.5. Visual description of NGC/IC galaxies (from Table 7.3 and Table 7.4) using different apertures; sorted by NGC/IC number (Local Group members are described in Table 8.2, IC 342 in Table 8.4)—Cont'd

NGC	Small/Medium aperture	Large aperture
4449	Faint, stellar (16 × 70). Bright, moderately large, elongated, bright core (SG 8″)	Very bright, very large, elongated SW–NE, bright core, stellar nucleus. A knot is involved at the N end and the galaxy generally appears brighter to the N of the core. A star is superimposed close E of the core (SG 17.5″)
4490	Faint spot (10 × 50). Bright, elongated, large core. Small companion (NGC 4485, round, dense patch) to the north (8″)	Very bright, striking, elongated 2:1 NW–SE, 6′ × 3′, large bright core is elongated and grainy. A faint arm extends from the NW end in the direction of NGC 4485 3.6′ NNW, a small extension (arm) at the SE end is suspected (SG 13″)
4494	Bright, fairly small, round, bright core (SG 8″)	Bright, pretty large, round, very suddenly very much brighter in the middle with a bright nucleus at 150×. This galaxy almost doubles in size with averted vision, it has a pretty high surface brightness (SG 17.5″)
4526	Bright, elongated halo, stellar core, between two stars (8″)	Very bright, fairly large, very elongated WNW–ESE, bright core, strong stellar nucleus. A 12.5 mag star is 1.3′ S of center (SG 17.5″)
4535	Bright, large halo with stars (8″)	Bright, fairly large, very small bright core, elongated SSW–NNE, about 5.5′ × 4.0′. Appears slightly darker on both sides of core (this is a gap between the spiral arms). A 13.5 mag star is superimposed on the N side 1.0′ from the center and a similar star is at the S end of the halo 2.2′ from center. A faint 14.5 mag star is just 48″ SW of the core (SG 17.5″)
4559	Faint, diffuse (16 × 70). Bright, irregular appearance, faint halo, two stars near (8″)	Bright, large, elongated 5:3 NW–SE, ~7′ × 3′. Exhibits a striking, unusual appearance with a broad, weak concentration to a large, elongated core. The overall surface brightness is noticeably irregular with hints of brighter and darker spots. The outer halo has a low surface brightness, particularly on the SE end which is cradled by three 12–12.5 mag stars. This end is wider than the NW side, shows no tapering and there appears to be mottling near the superimposed stars (SG 17.5″)
4565	Bright, very large, edge-on, bright center, dark lane barely visible (8″)	Bright, very large, edge-on 12:1 NW–SE, dimensions approximately 16′ × 1.5′. Beautiful dark lane visible continuously with direct vision along most of major axis although more prominent in the center. The galaxy is split asymmetrically by the dust lane with the southern half larger and brighter. There is subtle structure visible along the dust lane. Contains a small bright core with a stellar nucleus at the S edge of the lane. A 13.5 mag star is 1.6′ NE of the center (SG 17.5″)

4631	Easy, elongated (16 × 70). Large elongated; bright with well defined nucleus; northern edge sharp, southern edge more diffuse (8″, 100×)	Large edge-on galaxy; slightly asymmetric with bright nucleus and knots; southwest edge with cuts (16″, 200×)
4725	Visible in 16 × 70. Bright, round nucleus with small bar, diffuse halo (8″)	Very bright, impressive, very small bright core, elongated SW–NE, large halo. Structure is suspected with the WSW edge possibly brighter (SG 13″)
4753	Bright, small core, oval halo (8″)	Bright, large, oval 2:1 E–W, the halo brightens down to a small very bright core. Overall, an impressive galaxy (SG 17.5″)
4945	Very large and bright. A much elongated lens with a much brighter elongated core. Situated in a rich star field near a bright star (6″)	This long edge-on spiral is fairly bright and broadly concentrated with a slightly bulging core, extending SW–NE ~14′ × 2.5′. The surface brightness is relatively uniform with a weak central brightening and dimming toward the tips. Set in a rich star field peppered with faint stars (SG 18″)
5005	Faint, diffuse, brighter middle, nonstellar core (8″)	Very bright, large, elongated 5:2 WSW–ENE, 4.8′ × 2.0′. Strong concentration with a small very bright elongated core and stellar nucleus (SG 17.5″)
5068	Faint, round, low surface brightness (8″)	Fairly large, diffuse, no definite edges, almost round (SG 13″)
5102	Fairly faint, fairly large, elongated (SG 8″)	Pretty bright, pretty large, much brighter in the middle and elongated at 100×. Reminds me of a miniature M 31 (SG 17.5″)
5128	An extended area of haze surrounding an almost stellar center. Dark lane not seen at 32× but apparent at 70× when the object appears V-shaped (6″)	Bright, large, very large prominent dust lane oriented NW–SE. The SW hemisphere dominates in size and brightness. A star is superimposed at the S edge of the dust lane (W of center) and a bright star is superimposed on the SW hemisphere (S of center) (SG 17.5″)
5247	Faint, smooth halo, stellar core (8″)	Moderately bright, large, slightly elongated 4:3 SW–NE, about 4′ × 3′, sharp concentration with a very weakly concentrated halo which fades into the background. Unusual appearance as suddenly rises to very small bright core 20–30″ diameter. Spiral structure not seen (SG 17.5″)
5866	Bright, oval patch, bright center (8″)	Very bright, fairly large, elongated 2:1 NW–SE, 3.0′ × 1.5′, bulging bright core. This galaxy has a high surface brightness and a mottled surface. Just a hint of the razor-thin dust lane prominent on photographs is visible (SG 17.5″)
6946	Round, averted vision (16 × 70). Large, faint oval object of equal brightness; no structures; more difficult than M 101 (8″, 60×).	Bright, very large, 6′ diameter to main body, elongated 3:2 ~E–W. Three arms are visible. A long bright arm is attached at the N side of the core and trails to the E. This eastern arm splits; a short fainter branch bends S following the core and a long curving bright arm

(*Continued*)

Table 7.5. Visual description of NGC/IC galaxies (from Table 7.3 and Table 7.4) using different apertures; sorted by NGC/IC number (Local Group members are described in Table 8.2, IC 342 in Table 8.4)—Cont'd

NGC	Small/Medium aperture	Large aperture
		terminates with a very faint, very small HII knot. On the W side a fainter arm shoots sharply to the N from the core. These outer arms significantly increase the diameter of the main body. Has a very large brighter middle but the core is just a very small brighter region close to SW of the geometric center. A very faint stellar nucleus was seen with direct vision (SG 17.5″)
7331	Faint, oval (16 × 70). Pretty bright, pretty large, much elongated 3′ × 1′, an obvious edge on galaxy. 6.7 mm elongated 3′ × 1′ in PA 165°, the central region is bright and also elongated in the same PA as the galaxy. There is only one companion seen to the NE (SC 6″)	Very bright, very elongated 3:1 NNW–SSE, 9′ × 2.5′, very bright elongated core, substellar nucleus. The W side has a sharper edge due to dust (SG 13″). Core is three times seeing disk, dark lanes obvious. Seven companions seen, five above and two below, two of those held with averted vision only. Great view, almost entire field of view (SC 25″)
7793	Visible in 16 × 80 finder. Very large, oval, low surface brightness (SG 8″)	Bright, very large, oval 3:2 WSW–ENE, very large broadly brighter halo, small bright core (SG 17.5″)

Additional Notes

NGC 253. The largest and brightest member of the Sculptor Galaxy Group that also includes the bright galaxies NGC 55 (Fig. 1.12), NGC 247 (Fig. 7.6), NGC 300 (described below; Fig. 7.5), and NGC 7793. NGC 253 (Fig. 3.1) is a highly inclined, large dusty spiral that is undergoing a major star-forming episode in the region around the nucleus . Like M 82 (Fig. 1.21), this is an example of a nearby "starburst galaxy," and has a small, nearly stellar nucleus and a number of superstar clusters (SSCs).

NGC 300. A large "pinwheel" galaxy (Fig. 7.5) that resembles a smaller, fainter version of M 33 (Fig. 7.7) . Much like the later galaxy, it has two major and two less developed spiral arms that are well delineated by numerous giant HII regions and OB associations. Many of these structures are visible with larger telescopes when viewed from southern locations.

NGC 1023. The brightest member of a northerly group of galaxies that includes NGC 891 (Fig. 6.6) and NGC 925. NGC 1023A is a distorted companion located about 2.5′ east the main galaxy's core.

NGC 1097 and *NGC 1365.* These are two massive barred spirals that are members of the Fornax Galaxy Group. In addition to being "Grand Design Spirals," both are also Seyfert galaxies with brilliant nuclei. NGC 1365 is shown in Fig. 1.20. NGC 1097 is particularly interesting as it shows signs of distorting via interaction with nearby NGC 1097A, plus it has an unusually bright and well defined inner spiral pattern or nuclear ring.

Fig. 7.6. NGC 247 in Cetus, a member of the Sculptor group

NGC 1316. Sometimes called Fornax A, this is a bright giant elliptical that lies SW the core region of the Fornax Cluster. Deep exposures of this system reveal a series of dust lanes and arcs, indicative of a galactic cannibalism event nearly a billion years old. This merger of two galaxies is a more evolved version than that associated with NGC 5128 (Fig. 7.8).

NGC 1532/1. A spectacular example of an edge-on "Whirlpool" class interacting galaxy system (Fig. 7.3) [169]. The larger galaxy (NGC 1532) has a large dust lane that cuts across the bright disk and the smaller object has warped the SW end. On most images, the tidal tails can be seen stretching toward the smaller galaxy.

NGC 2841. This galaxy is an archetype of the "flocculent" spiral arm class. Instead of having large, well defined arms – they are a series of short segments. Another well-known example of this class is the "sunflower galaxy" M 63. This galaxy has also produced several bright supernovae over the past fifty years.

NGC 2903. One of the brightest of all non-Messier galaxies, this large barred spiral has considerable star formation activity near the nucleus. Called a "hotspot" galaxy, most of the star formation is associated with a circumnuclear ring-like structure surrounding the system's core [222].

NGC 3115. The "Spindle Galaxy," this is perhaps the finest example of a lenticular or SO galaxy in the sky.

NGC 3628. With M 65 and M 66, this forms the "Leo Triplet." A large edge-on system with a massive, scalloped dust lane that is visible in modest telescopes under dark skies (Fig. 7.4).

NGC 4214 and *NGC 4449.* Both systems are magellanic irregulars undergoing massive starburst activity. Each has numerous large OB associations and HII regions that can

Fig. 7.7. M 33 in Triangulum with the bright HII region NGC 604 (at east end of the northern arm)

be readily resolved in larger telescopes. NGC 4449 in particular has a high surface brightness allowing for high magnification scanning of its complex structure Fig. 7.9). The stellar "nucleus" of NGC 4449 is an extremely luminous example of a SSC.

NGC 4490/85. A bright pair of interacting galaxies with elongated tidal plumes. NGC 4490 has been warped by the encounter, with an elongated plume being drawn toward the more compact NGC 4485 [169]. The unusual shape of this galaxy has lent to the popular nickname the "Cocoon Nebula" (not to confuse with the galactic nebula IC 5146).

NGC 4565. Perhaps the finest edge-on galaxy in the sky, it is impressive in almost any sized instrument (Fig. 1.14). The disk stretches over 0.25°, or at the currently accepted distance – over 150,000 ly.

Fig. 7.8. The peculiar radio galaxy Centaurus A (NGC 5128)

NGC 4631/27. Another gigantic edge-on galaxy whose disk has been warped by the interaction of the nearby dwarf galaxy NGC 4627 (Fig. 3.8). Unlike the more famous NGC 4565, it has an irregular, broken dust lane and numerous large HII regions and associations. About 30′ to the SE lies NGC 4656/7, better known as the "Hockey Stick." It is a highly disturbed system, as a result of tidal interactions with NGC 4631 in the distant past.

NGC 5128. Known as Centaurus A, this a giant radio galaxy and largest member of group that includes NGC 4945, NGC 5102, and M 83 (Fig. 3.3). Its most noticeable feature is massive dust lane that bisects the galaxy in half (Fig. 7.8). Recent data from the HST have revealed that this broad irregular band is the remains of a smaller spiral galaxy that had collided with a huge elliptical galaxy several hundred million years ago (see Waller & Hodge). This is a classic example of what is known as "galactic cannibalism" and in time the entire galaxy will be "consumed." A possible fragment of this collision is ESO 270-17, a thin "shard" lying around 3′ to the SE of the main system. Measuring 15.5′ × 1.4′, it requires large aperture under dark skies for observation.

NGC 5866. A beautiful lenticular galaxy that is presented almost perfectly edge-on. It has a narrow, razor-sharp dust lane that is visible in larger scopes. This object has been the center of controversy for over 150 years since Admiral Smyth's claim that this galaxy was originally discovered by Pierre Mechain. Now called "Mechain's Lost Galaxy," it is considered here – like many other authors do – as Messier's missing galaxy M 102.

NGC 6946. A large, nearly face-on spiral galaxy lying near the Cygnus–Cepheus border (Fig. 1.16). This galaxy is the site of intense starburst activity and is noted for its unusually high incidence of supernovae. The four spiral arms and numerous HII regions may be glimpsed with larger telescopes under better skies.

NGC 7331. A large, highly inclined spiral galaxy (Fig. 7.10) discovered by William Herschel in 1784. Lord Rosse first noted the beautiful spiral structure in his original list of 14 "spiral nebulae." It is surrounded by a host of smaller galaxies, though most of these are background objects.

Bright UGC Galaxies

Representing the many general catalogues of galaxies, we feature the *Uppsala General Catalogue* (UGC) and its extension (UGCA). It contains (nominally) bright galaxies not

Fig. 7.9. NGC 4449 in Canes Venatici, an irregular starburst galaxy with many bright knots

Fig. 7.10. NGC 7331 in Pegasus and its background galaxies

listed in the NGC/IC, e.g., UGC 3714 in Camelopardalis (Fig. 7.11). Some show a pretty low surface brightness. Nevertheless, considering values of V and V′, which promise a visibility with medium aperture; we get a sample of 16 galaxies (Table 7.6; visual descriptions in Table 7.7).

Additional Notes

UGC 2885. This is the largest known Sc spiral galaxy, measuring over 800,000 ly is diameter, or nearly ten times the size of our own Milky Way galaxy! It weighs in at over a trillion solar masses and has a rotation period of over two billion years.

UGC 3697. The Integral-Sign Galaxy. Its unusually shape derives from tidal interaction with nearby UGC 3714, which has warped the tips of the galaxy's disk (Fig. 7.11).

Fig. 7.11. Pretty bright UGC 3714 in Camelopardalis and its strange companion, the "Integral-Sign Galaxy" UGC 3697

Table 7.6. Bright galaxies from UGC and UGCA ($V \leq 12.2$ mag and $V' \leq 14.5$ mag/arcmin^2)

UGC(A)	Con	R.A.	Decl	V	V'	$a \times b$	PA	Type	Remarks
UGC 1886	And	02 26 00.6	+39 28 13	11.9	14.0	3.7 × 2.0	35	Sb	
UGC 2885	Per	03 53 02.5	+35 35 23	12.8	14.4	3.2 × 1.6	40	SA (rs)	Largest known spiral
UGCA 127	Mon	06 20 55.6	−08 29 41	12.2	13.6	3.9 × 1.1	70	Scd?	
UGC 3580	Cam	06 55 31.0	−69 33 45	11.8	13.7	3.4 × 1.9	3	Sa	
UGC 3685	Lyn	07 09 05.7	+61 35 44	12.0	13.3	2.0 × 1.9	150	SBb	
UGC 3691	Gem	07 08 01.3	+15 10 45	11.9	12.6	2.2 × 1.0	65	Sc	
UGC 3714	Cam	07 12 32.8	+71 45 00	11.9	12.8	1.8 × 1.5	35	S? pec	Near UGC 3697 ("Integral-Sign Galaxy")
UGC 3972	Cam	07 44 42.2	+73 49 16	11.8	13.4	1.2 × 0.7	160	SB pec	Arp 17
UGC 8041	Vir	12 55 12.7	+00 06 59	12.0	13.7	3.0 × 1.8	165	SBd	
UGC 8287	Cam	13 11 02.9	+78 24 45	11.8	13.7	1.3 × 0.8	155	SBa	
UGC 9748	UMi	15 07 23.0	+76 02 56	11.8	14.0	1.4 × 1.0	20	SB0	Pair w. UGC 9750
UGC 11453	Cyg	19 31 08.2	+54 06 04	12.0	12.7	1.7 × 1.3	62	Sb	Pair w. PGC 63313
UGC 11466	Cyg	19 42 58.5	+45 17 51	11.8	12.3	1.7 × 1.1	35	S0? pec	
UGC 11781	Cyg	21 36 39.3	+35 41 40	12.1	12.3	1.4 × 1.0	75	SAB0	
UGC 11920	Lac	22 08 27.5	+48 26 27	11.9	13.1	2.3 × 1.5	45	SB0/a	
UGC 11973	Lac	22 16 49.8	+41 30 00	12.1	13.0	3.0 × 0.9	42	SBc	

Table 7.7. Visual descriptions of bright UGC(A) galaxies

UGC	Description
1886	Bright core, faint halo, slightly oval (14″). At 280× easily visible as a faint, moderately large glow, elongated 3:2 SW–NE, ~1.0′×0.6′. Fairly even concentration to a small brighter core. The outer extent of the faint halo increases in size with averted vision (SG 17.5″)
2885	Fairly large, diffuse, elongated NE–SW oval with a nearly stellar core (RJ 20″)
3580	Fairly faint, small, diffuse. A faint star is off the E end 25 in. from the center. There is a larger very faint halo at low power but still appears smaller than the listed size (SG 13″)
3685	Faint, fairly small, elongated 4:3 SW–NE, 0.6′×0.45′. A faint star is on the western edge (SG 17.5″)
3691	Faint, moderately large, elongated 3:2 N–S, fairly low even surface brightness, no central concentration. Appears similar to a faint nebulosity in a rich milky way field. A 10 mag star at the NW edge 1.0′ from center which detracts from viewing. A 12 mag star is at the N edge 44 in. from the center and a fainter 13 mag star is at the S edge a similar distance from center (SG 17.5″)
3714	Fairly bright, round, high surface brightness, UGC 3697 ("Integral-Sign Galaxy") near (14″). Moderately bright, round, bright core. This galaxy has a surprisingly high surface brightness for a UGC galaxy (SG 17.5″)
3972	Pretty bright, oval, stellar nucleus (14″)
8041	Faint, very large, elongated NNW–SSE. Has a low irregular surface brightness with some brighter portions (SG 17.5″)
8287	Fairly bright, small, oval, stellar core (14″)
9748	Fairly bright, small, round, stellar core (14″)
11453	Fairly bright, large, slightly elongated, small center (20″)
11466	Moderately bright, elongated 3:2 SW–NE, 1.3′×0.8′, broad very weak concentration, tapers toward the SW end (SG 17.5″)
11781	Faint, very small, round, 20″ diameter, very weak even concentration. A 13 mag star lies 1.2′ NE (SG 17.5″)
11920	Fairly faint, small, round. Dominated by a bright 30″ core with a much fainter low surface brightness halo with averted vision. The core increases to an occasional stellar nucleus. Very difficult to determine outer extent as quickly fades into background (SG 17.5″)
11973	Fairly faint, moderately large, very elongated 3:1 SW–NE, large brighter middle, fairly low surface brightness (SG 17.5″)
UGCA127	Fairly bright, large, elongated, small center, faint halo (20″)

Sky Areas and Constellations

Instead of scanning a catalogue, you may scan a certain area of the sky for galaxies. The area can be defined by a star pattern (large asterism), a constellation, or part of within it. From there, you can compile all galaxies in this area that are within the reach of your telescope. But be careful: owners of a 20″, you should not start with Virgo! Then its time to start the process of "galaxy hopping" with a small instrument. For instance, you can observe all brighter objects in Leo with a 6″ [127].

Promising areas for 8–12″ telescopes are the Pegasus square [128], the bowl of the Big Dipper, the body of Hercules or constellations like Canes Venatici [129,130], Corona Borealis, and Triangulum [131]. Even a small, inconspicuous constellation like Leo Minor

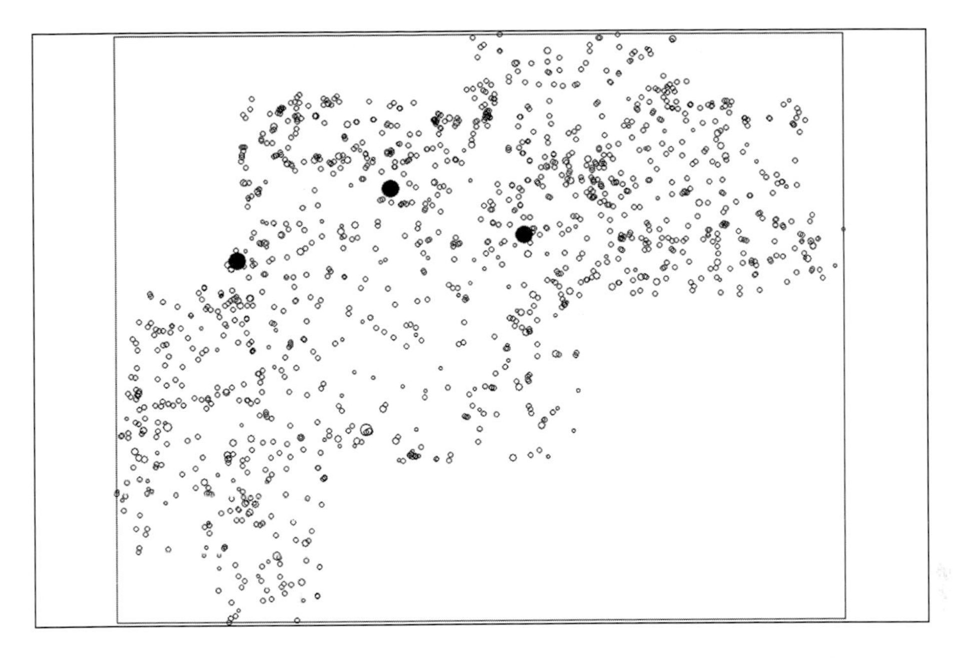

Fig. 7.12. Distribution of 1,400 galaxies in Leo Minor

can be an overwhelming task when considering all of the 20 "targets (Fig. 7.12) – over 1400 galaxies create an "Amateur Deep Field" [132]! The major galaxies are presented in [250]. A better choice might be tiny Equuleus [133]. Finally, how about the galaxies located within the Milky Way's "zone of avoidance"? With medium sized apertures you can find many fine objects [134]. Constellations such as Cygnus [135], Orion [136], Pegasus [137], or Lyra [138] are nice places to look too.

Galaxies Near the Celestial Poles

There are two special areas in the sky that are not emphasized by star patterns: the celestial poles [139]. Sweeping around them is perhaps the easiest with an altazimuth-mounted telescope. You will find a lot of galaxies within 5° of either pole (Tables 7.8 and 7.9). The most prominent northern object is NGC 3172, better known as "Polarissima Borealis" (Fig. 7.13). It is visible in a 12″, whereas its small companion MCG 15-1-10 needs at least 16″ aperture. The southern counterpart, NGC 2573 in Octans ("Polarissima Australis;" Fig. 7.14), is a bit brighter. Visual descriptions are given in Table 7.10.

Table 7.8. A selection of galaxies within 5° of the northern celestial pole (sorted by decreasing declination)

Object	Con	R.A.	Decl	V	V′	a × b	PA	Type	Remarks
NGC 3172	UMi	11 47 14.5	+89 05 37	14.4	13.9	1.0 × 0.7	39	Sb	Polarissima Borealis
MCG 15-1-10	UMi	11 40 40.2	+89 05 06	15.2	15.3	1.1 × 1.0		E?	Pair w. NGC 3172 (sep. 4′)
UGC 10923	UMi	17 19 32.4	+86 44 08	13.4	12.9	1.2 × 0.6	6	S?	VV 706 (pair)
UGC 10740	UMi	16 53 28.5	+86 35 25	14.4	14.0	1.0 × 0.8	140	SB0	Akn 518
UGC 3536A	Cep	07 03 22.2	+86 33 26	13.6	12.7	0.7 × 0.6		E?	Arp 96, pair w. UGC 3528A
NGC 1544	Cep	05 02 36.2	+86 13 22	13.6	13.6	1.3 × 0.9	130	S?	
NGC 2276	Cep	07 27 13.8	+85 45 18	11.3	12.8	2.3 × 1.9	20	SAB(rs)c	Arp 25, Arp 114
IC 499	Cam	08 45 17.3	+85 44 26	13.0	13.8	2.1 × 1.1	80	Sa	
UGC 3654	Cep	07 17 46.7	+85 42 44	14.3	11.8	0.4 × 0.3	26	S0 compact	
NGC 2300	Cep	07 32 20.3	+85 42 33	11.1	13.0	2.8 × 2.0	78	SA0	Arp 114
IC 455	Cep	07 34 58.7	+85 32 16	13.3	12.9	1.1 × 0.7	69	S0	
IC 512	Cam	09 03 49.0	+85 30 06	12.3	13.1	1.8 × 1.3	1	SAB(s)cd	
UGC 1198	Cep	01 49 17.8	+85 15 36	13.8	13.1	0.8 × 0.6	85	E?	VII Zw 3
IC 469	Cep	07 55 59.3	+85 09 31	12.7	13.4	2.2 × 1.0	90	SAB(rs)ab:	

Table 7.9. Galaxies within 5° of the southern celestial pole (sorted by increasing declination)

Object	Con	R.A.	Decl	V	V'	$a \times b$	PA	Type	Remarks
NGC 2573	Oct	01 41 53.2	−89 20 03	13.4	13.6	1.9 × 0.7	85	SAB(s)cd:	Polarisima Australis
NGC 2573A	Oct	23 12 14.6	−89 07 27	13.9	14.0	2.1 × 0.6	18	SBb? pec	
NGC 2573B	Oct	23 07 32.0	−89 06 55	14.5	14.0	1.5 × 0.5	120	IBm? pec	
ESO 8-8	Oct	15 31 18.5	−87 26 05	13.3	13.3	2.3 × 0.5	175	SB(s)dm sp	
ESO 6-6	Oct	09 15 42.8	−86 46 02	13.4	13.0	0.9 × 0.7	179	E2:	
ESO 5-4	Oct	06 05 39.7	−86 37 54	12.4	13.2	3.9 × 0.6	97	Sb: sp	
ESO 4-20	Oct	05 29 24.8	−85 55 46	13.9	13.0	1.0 × 0.5	76	Sb	
ESO 2-12	Oct	01 00 50.0	−85 31 22	14.3	13.0	1.8 × 0.2	49	Scd? sp	RFGC 233
ESO 5-11	Oct	07 42 03.5	−85 25 21	13.7	12.7	1.2 × 0.4	110	S0-sp	
NGC 6438A	Oct	18 22 33.0	−85 24 22	11.6	12.5	2.7 × 1.0	32	Ring galaxy	
NGC 6438	Oct	18 22 15.9	−85 24 06	11.7	12.5	1.6 × 1.4	156	Ring galaxy	
ESO 9-10	Oct	17 39 32.1	−85 18 34	11.9	13.0	2.1 × 1.5	171	SA(s)bc:	
ESO 3-1	Oct	01 32 00.3	−85 10 41	12.8	13.1	1.3 × 1.2	0	(R)SB(rs)0/a	

Fig. 7.13. "Polarissima Borealis" NGC 3172 in Ursa Minor

Fig. 7.14. "Polarissima Australis" NGC 2573 in Octans

Table 7.10. Observations of galaxies near the celestial poles (sorted by designation)	
Galaxy	Description
NGC 1544	Fairly faint, small, round. Several faint stars are near including an evenly matched 14.5 mag pair with 10′ separation at the N edge 20″ from center (SG 17.5″)
NGC 2276	Faint, moderately large, low surface brightness, slightly elongated. A 9 mag star is near (SG 8″). Diffuse, slightly elongated. Located 2.2′ ENE of 8.4 mag star which interferes with viewing. Forms a pair with NGC 2300 6.4′ ESE (SG 13″)
NGC 2300	Moderately bright, small, bright core, slightly elongated (SG 8″). Fairly bright, bright core, small fainter halo. Forms a pair with NGC 2276 7′ W (SG 13″)
NGC 2573	Small slightly elongated puff of light, with low surface brightness, brightening slightly toward the center core. The western edge of the galaxy seems to be a little more hazy. A string of evenly spaced stars run out in a half moon shape started just off the galaxy western hazy edge (Magda Streicher 12″). Faint 2′ × 0.5′ haze, elongated ENE–WSW, which rises only slightly to the middle and has a bright mag 15 stellar nucleus or superposed star. The W end of the disk is slightly brighter and broader and averted vision shows a large faint round brightening W of the nucleus. Two 14/15 mag stars are 1.5′ W. Best viewed at 450× (Michael Kerr 25″). Small faint elongated smudge brightening to the center, with no obvious suggestion of spiral structure. Field star involved, perhaps accounting for the central brightening (Lynton Hemer 30″)
NGC 2573A/B	Both the galaxies situated close to one another 30″ from Polarissima to the west. I could not confirm NGC 2573A and NGC 2573B (Magda Streicher 12″)
NGC 3172	Faint, small, round (14″). Faint, small, round, brighter core, faint stellar nucleus, can hold steadily with averted vision. A 12.5 mag star is 1.5′ distant (SG 17.5″)
NGC 6438/A	Quite interesting. Best viewed at 450×. NGC 6438 appears as a bright round 0.5′ diameter haze, which brightens to a nonstellar core. NGC 6438A is attached on the SE side and appears initially as an irregular 0.5′ diameter haze. Averted vision shows it to be extended to the NE with an overall size of 1′ × 1.5′. The NW edge of NGC 6438A is brighter and mottled and a faint extension of this edge to the SSW is also just visible (Michael Kerr 25″)
IC 455	Fairly bright, small, oval, compact core (20″)
IC 469	Fairly faint, elongated, smooth disk, low surface brightness (20″)
IC 499	Pretty bright, small, oval (20″)
IC 512	Fairly faint, moderately large, round, almost even surface brightness. A 10′ string of stars just E is oriented roughly N–S with a 9 mag star at the N end (SG 17.5″)
UGC 1198	Faint, round, compact core with small halo (20″)
UGC 4297	Faint, pretty flat, compact core (20″)
UGC 10923	Faint, irregular round (20″)
MCG 15-1-10	Very faint, small, round, NGC 3172 near and much easier (20″)
ESO 2-11	Very faint 0.8′ × 0.5′ haze, elongated NE–SW, which rises only slightly to the middle. Visible with direct vision but best seen with averted. A mag 15 star is on the SW end, a pair of 14 mag stars is 2′ SSW and a 14 mag star is 1.8′ NNE. Best viewed at 350× (Michael Kerr 25″)

(*Continued*)

Table 7.10. Observations of galaxies near the celestial poles (sorted by designation)—Cont'd

Galaxy	Description
ESO 2-12	Very faint 1.4′ × 0.2′ haze, elongated NE–SW, which is really only seen with averted vision. A 14 mag star is 2.2′ SSW and a 13 mag star is 2.5′ ESE. Best viewed at 350× (Michael Kerr 25″)
ESO 3-1	Faint 1′ × 0.5′ haze, elongated N–S, which rises sharply to a small core and bright 14 mag stellar nucleus. The central region shows two small slightly brighter patches 0.3′ N and S of the core. A 12 mag star is 3′ SSE. Best viewed at 350× (Michael Kerr 25″)
ESO 4-20	Faint 1′ × 0.5′ slightly grainy haze, elongated E–W, which rises broadly to the middle and shows glimpses of a faint stellar nucleus with averted vision. A pair of 15/16 mag stars is 2.3′ ESE and a small arc of three 14/15 mag stars is 3′ NE (Michael Kerr 25″)
ESO 5-4	With stars in the shape of an arrow indicating direction, I could identify the position quite easily. A faint glow with a characteristic oblong appearance (W/E) which blends into the background. Not easy to observe the very faint galaxies and usually I have to make use of averted vision. ESO 5-4 was however, in a way fairly easy (Magda Streicher 12″). Faint 3′ × 0.5′ streak, elongated E–W, which rises from very faint extremities to a brighter 1′-long central region and stellar nucleus. The N side of the streak is sharply cut off by a dust lane and the S side is flat and also appears fairly sharply cut off. With averted vision the dust lane appears as an actual lane, which means there must be an extremely faint part of the streak on the N side of the lane. Bracketed by a 14 mag star 0.7′ N and a 16 mag star 0.6′ S with a 12 mag star 2.3′ SSW. Best viewed at 350× (Michael Kerr 25″)
ESO 5-11	A good directive is provided by the two stars on the western side situated in line with the galaxy. After a long battle, averted vision and cloth over my head, I could identify a very faint source of light which appears slightly round in shape (Magda Streicher 12″). Fairly faint 0.8′ × 0.4′ haze, elongated ESE–WNW, which rises to a small core and bright stellar nucleus. A 15/16 mag star is superposed 8″ W of the nucleus. Forms a triangle with two 14/15 mag stars 1.5′ S and a 9 mag star is 2.5′ W. Best viewed at 450× although 650× shows the small core better (Michael Kerr 25″)
ESO 6-6	Fairly faint 0.5′ × 0.3′ haze, elongated N–S, which rises to a stellar nucleus. The rise in surface brightness is not completely smooth in the centre suggesting some structure but nothing definite can be seen even at 450×. A 12 mag star is 0.9′ E and a 14 mag star is 1.2′ E. Best viewed at 350× (Michael Kerr 25″)
ESO 8-8	Faint 2′ × 0.4′ haze, elongated N–S, which rises only slightly to the middle. The disk is slightly broader on the N side and appears slightly convex to the E. Bracketed by a five 14/15 mag stars. Best viewed at 350× (Michael Kerr 25″)
ESO 9-10	Five well-placed stars indicate the position of the faint galaxy. Round to fairly large and very hazy with a slightly brighter centre which looks like a faint little star out of focus. With a low surface brightness, the contrast between the haziness and the darker background is improved with averted vision (Magda Streicher 12″). Moderately bright 2.5′ × 1.7′ haze, elongated N–S, which is slightly mottled and rises from a diffuse periphery to a small elongated core and stellar nucleus. A mag 16 star is superposed 0.8′ SW and a 14 mag star is 1.5′ NE. Best viewed at 350× (Michael Kerr 25″)

Chapter 8

Individual Objects

Let us now use selection criteria that are dependent on the characteristics of the individual galaxy. We should naturally start with apparent brightness and size. But this aspect is implicitly contained in the previous lists, including the Messier galaxies and the brightest NGC/IC- and UGC galaxies. We therefore will concentrate on distance, which is not always synonymous with brightness, as the dwarfs or quasars demonstrate. Later the appearance of galaxies will be featured, such as the degree of elongation (e.g., edge-on galaxies) and peculiar and/or unusual structures.

Small Distance: Nearby Galaxies, Dwarfs, Associated Non-Stellar Objects

Sorting objects by distance we naturally start with the nearest: members of the Local Group and galaxies that lie just beyond [140,141]. Many of these nearby systems are dwarfs. If there is a chance of visual observing such faint systems at all, it is here. Interesting features of nearby galaxies include a variety of nonstellar objects: bright HII regions, OB associations, super star clusters (SSC), and globular clusters (GC).

Galaxies of the Local Group (LG) and Beyond

The first conspicuous galaxies we meet outside the Milky Way are the Large and Small Magellanic Clouds, lying 163,000 ly and 196,000 ly, respectively, away from our solar system. Though both are visually impressive and bright, they are relatively small galaxies. The nearest really large systems are M 31 and M 33, lying at distances of 2.5 million ly and 3.0 million ly, respectively. M 31 is a huge Sb (or perhaps SBb) spiral galaxy that is considerably larger and more massive than our own Milky Way. It hosts over 300 globular clusters, plus numerous open clusters and OB associations – many of which are visible in amateur scopes. M 33 is a much smaller Sc spiral, with two prominent spiral arms (plus two smaller fragmented ones) that are peppered with huge HII regions. Spanning at most 60,000 ly and with less than 10% of the mass of our galaxy, its not much larger than the Large Magellanic Cloud [207].

But there is a lot more to see within 1.2 Mpc (3.9 million ly), which is defined as the limit of the Local Group (Table 8.1). Currently there are nearly 40 confirmed members of the LG, of which more than 50% are dwarf irregulars and spheroids. Visually LG members can range from easy (naked eye) to extremely challenging (20″ and up), or simply invisible, as in the case of the Sagittarius dwarf elliptical (SagDEG), being tidally disrupted on

Table 8.1. Local Group galaxies beyond the Milky Way with a (working) distance of 1.2 Mpc (Dist = distance from the Milky Way center in kpc)

Galaxy	Con	R.A.	Decl	V	V'	$a \times b$	Type	Dist	Remarks
WLM	Cet	00 01 58.0	–15 26 59	10.6	14.1	9.5×3.0	IB(s)m	950	DDO 221
IC 10	Cas	00 20 24.5	+59 17 33	11.2	14.9	6.4×5.3	IB(s)m	660	
Cetus	Cet	00 26 10.8	–11 03 14	14.0	15.4	6×4	dSph (E4)	780	
NGC 147	Cas	00 33 11.7	+48 30 26	9.4	14.5	13.2×7.8	dE5	710	DDO 3
And III	And	00 35 31.3	+36 30 31	14.0	15.2	4.5×3.0	dSph (E3)	750	
NGC 185	Cas	00 38 57.6	+48 20 14	9.3	13.7	8.0×7.0	dE3	640	
M 110	And	00 40 22.1	+41 41 07	7.9	13.8	19.5×11.5	SA0-	810	NGC 205
And VIII	And	00 42 18.0	+40 37 00	9.1	11.5	45×10	dSph (E pec)	770	
M 32	And	00 42 41.8	+40 51 57	8.1	12.5	8.5×6.5	cE2	770	NGC 221
M 31	And	00 42 44.3	+41 16 08	3.5	13.5	189.1×61.7	SA(s)b	770	NGC 224
And I	And	00 45 41.5	+38 02 09	12.6	15.4	4.0×3.0	dSph (E3 pec?)	810	
SMC	Tuc	00 52 40.0	–72 48 34	2.2	14.1	319×205	IB(s)m	60	NGC 292
And IX	And	00 52 53.0	+43 11 45	16.2	17.0	5.9×1.3	dSph	735	
Sculptor	Scl	01 00 09.3	–33 42 37	10.0	18.9	40.0×31.0	dSph (E3 pec)	85	ESO 351-30
Pisces	Psc	01 03 54.0	+21 53 00	18	19	2×2	IAm	620	LGS 3
IC 1613	Cet	01 04 54.2	+02 08 02	9.3	15.1	16.6×14.9	IB(s)m	740	DDO 8
And V	And	01 10 17.2	+47 37 41	15	15.5	2×1	dSph	810	
And II	Psc	01 16 26.3	+33 25 37	12.5	14.1	3.6×2.5	dSph (E0)	680	
M 33	Tri	01 33 51.9	+30 39 29	5.5	14.0	68.7×41.6	SA(s)cd	930	NGC 598
Phoenix	Phe	01 51 06.3	–44 26 51	12.4	14.7	4.9×4.1	IAm	440	
Fornax	For	02 39 59.0	–34 27 00	8.0	15.6	17.0×13.0	dSph (E2)	140	ESO 356-4
LMC	Dor	05 23 17.8	–69 45 21	0.4	14.1	650×550	IB(s)m	50	
Carina	Car	06 41 36.7	–50 57 58	16.8	19.3	22.0×15.0	dSph (E3)	100	ESO 206-220
Canis Major	CMa	07 15	–27 30			900×300	dE?	13	
Leo A	Leo	09 59 25.9	+30 44 43	12.6	15.5	5.1×3.1	IBm	800	DDO 69, Leo III
Leo I	Leo	10 08 28.5	+12 18 18	10.5	15.2	9.8×7.4	dE3	260	DDO 74
Sextans	Sex	10 13 02.9	–01 36 53	12	19	90×65	dSph (E3)	85	

Leo II	Leo	11 13 27.4	+22 09 39	12.0	17.0	10.1×9.0	dSph (E0 pec)	210	DDO 93, Leo B
Ursa Minor	UMi	15 08 48.5	+67 11 33	10.9	17.8	30.0×19.0	dSph (E4)	75	DDO 199
Draco	Dra	17 20 12.6	+57 55 05	9.9	17.9	50.0×31.0	dSph (E0 pec)	95	DDO 228
SagDEG	Sgr	18 55	–30 28	3.1	8.0	360×120	dSph (E7)	25	
SagDIG	Sgr	19 29 59.6	–17 40 42	13.2	14.6	2.9×2.1	IB(s)m	1200	
NGC 6822	Sgr	19 44 56.6	–14 48 23	8.7	14.4	15.4×14.2	IB(s)m	500	Barnard's Galaxy
Aquarius	Aqr	20 46 51.8	–12 50 53	13.9	14.7	2.2×1.1	IB(s)m	950	DDO 210
Tucana	Tuc	22 41 49.6	–64 25 11	11.8	13.2	2.9×1.2	dSph (E5)	900	
And VII	Cas	23 26 31.0	+50 41 31	13	14	2.5×2.0	dSph	690	Cassiopeia Dwarf
Pegasus	Peg	23 28 36.4	+14 44 24	12.6	13.8	5.0×2.7	IAm	760	DDO 216
And VI	Peg	23 51 46.0	+24 35 00	14	15	4.0×2.0	dSph	790	Pegasus Dwarf

the opposite side of the Milky Way. Not included in the table in the newly found dwarf in Ursa Major at a distance 330,000 ly and with an extremely low absolute magnitude of -6.75 mag [234]. Visual descriptions of some "brighter" objects are given in Table 8.2.

You may ask why And is IV missing in the list? This is actually a star cloud in the outer disk of M 31 and not a proper galaxy.

Many of the Local Group galaxies are readily visible in small telescopes, but there are a number are challenging objects [142,143]. Pretty easy are NGC 147 and NGC 185 (Fig. 8.1), dwarf elliptical companions of M 31. Far fainter and more elusive are dwarf spheroidal systems, like Sculptor, Fornax, Ursa Minor, and Draco. Please note that most of these dwarfs can only be seen with a richfield telescope using low magnification [144]. Under exceptional sky conditions you get the impression of a large, dim glow, created by the sum of the many faint stars. In a larger telescope, these objects "disappear" due to the lack of contrast and are just too faint to be visible. This is also true for individual stars.

Additional Notes

NGC 147 and *NGC 185*. Along with M 110 (NGC 205) these are companions of the massive Andromeda Nebula (NGC 185 is shown in Fig. 8.1). Both systems have proven remarkably difficult to classify – as they have been designated as "dwarf ellipticals," "spheroids," and "dwarf spheroids" over the past twenty years [207]. Spheroids and dwarf spheroids are the most common type of galaxy in the universe and comprise well over 50% of the LG. In general, all three are strongly elongated and host small collections of globular clusters (GCs) – many of which are visible to observers with large telescopes.

NGC 6822. A dwarf barred irregular that is best known as "Barnard's Galaxy." Discovered visually by the famed 19th century comet hunter/celestial photographer E.E. Barnard with a 6 in. rich-field refractor, this object would prove to be extremely difficult to observe in the large, long focus refractors of the day. Later on, the study of Cepheid variables in this system (along with M 31 and M 33) was pivotal in Hubble's (1925) first determinations of extragalactic distances [207]. Numerous OB associations and HII regions are visible strewn along the bright bar and along the northern end of the galaxy, e.g., IC 1308 (Fig. 4.3). NGC 6822 is also one of the very few external systems that can be resolved into stars with a large to very large instrument. The brightest supergiants have a magnitude of ~16.5 mag, and under moderate magnification the galaxy can take on a granular look as the bar stars are being resolved.

IC 10. A highly obscured object that Hubble once called "one of the most curious objects in the sky" [207]. Detailed studies have revealed a high star formation rate and an unusually large number of Wolf-Rayet stars. Classified as a dwarf irregular, it is also the only starburst galaxy in the LG.

IC 1613. Originally discovered by the German astrophotographer Max Wolf back in 1906 [207]. This is a more "typical" dwarf irregular with only modest regions of star formation.

Fornax System. Discovered (along the Sculptor System) by Mrs. Lindsay from a photographic survey completed in 1938, this is the brightest "dwarf spheroid" (dSph) of the LG [207]. What sets this system apart from the other representatives in the LG is its population of globular clusters. All five GCs can be detected in a large telescope, while the brightest (NGC 1049; Fig. 8.2) is visible in fairly small instruments [154]. The galaxy itself is much more difficult, sometimes completely invisible in large

Table 8.2. Observations of Local Group galaxies (Messier objects were already described in Table 7.2)

Galaxy	Description
NGC 147	Very faint, moderately large, slightly elongated, diffuse (SG 8″). Fairly faint, very large, elongated almost 2:1 SSW–NNE, 5′×3′, very low almost even surface brightness. Contains a faint stellar nucleus or a 13.5 mag star is superimposed just N of center. Gradually fades into background (SG 17.5″)
NGC 185	Fairly faint, fairly large, diffuse (SG 8″). Bright, very large, slightly elongated ~E–W, broad concentration but no nucleus. Three 14 mag stars are at the W, NW, and SW ends. Higher surface brightness than NGC 147 (SG 17.5″)
NGC 6822	Very faint, elongated N–S (SG 8″). Fairly faint, very large, low but uneven surface brightness, elongated 5:2 N–S, 14′×6′. Diffuse appearance and the boundary is difficult to define, requires low power. Several faint stars are superimposed with a couple of brighter stars on the N side (SG 17.5″, 82×)
IC 10	Very faint, moderately large, elongated NW–SE. Unusually low even surface brightness. A 13 mag star is superimposed near the center. Located in a very rich star field (SG 13″)
IC 1613	Very faint, pretty large, very, very little brighter middle, elongated 1.5′× 1′ in PA 60°. Low surface brightness (SC 6″). Fairly faint, small, slightly elongated 4:3 NW–SE, very small bright core, stellar nucleus (SG 17.5″)
And I	4′ diameter, round glow of uniform brightness, just above the brightness level of the night sky (Tom Polakis 13″)
And II	Extremely faint, pretty large, not brighter in the middle, irregularly round, somewhat mottled at 100×. Rocking the tube of the scope helps the contrast with this very low surface brightness object (SC 13.1″)
And VI	Appears 8′×3′ in extent, elongated in PA 105°, and pretty faint overall but with some central brightening. There are 8 stars across the face of the galaxy in what is otherwise a poor starfield (Tom Polakis 13″). A very weak glow spotted about 1.5′ N of a 10 mag star. The main body of the galaxy roughly framed by a several 14–15 mag stars (RJ 20″)
And VII	Suspected only, uncertain of 'true detection' due to a scattering of 11–14 mag stars on the western limb of the object. The view seemed more certain at the lower magnification, however the light scattering from the foreground stars detracted considerable at increased magnification (RJ 20″)
Draco	Very faint, very large, elongated 1.5′×1′ in PA 110° at 60×. There are 10 stars involved across the face of the galaxy. I do not know if they are truly members of this nearby Local Group Galaxy. It is just a grainy lump at even this very low power and I was using a dark cloth (the monk's hood) over my head at an excellent site on a night rated at 8/10 for transparency. I was using a 38 mm eyepiece in a 2 in. barrel. So, if you are going to chase this very low surface brightness object, put in your lowest power (SC 13.1″). In excellent transparency it appeared as a very low surface brightness glow at 100×, roughly 15′×10′, elongated N–S. On the eastern side are a couple of 11 mag pairs and the glow extends just beyond a N–S string of stars on the west side. There appears to be a locally brighter region (possibly the core) offset toward the south side of the glow. The edges of the halo are difficult to follow as it generally fades into the background, though some areas seem to have a more well-defined edge (SG 17.5″)

(*Continued*)

Table 8.2. Observations of Local Group galaxies (Messier objects were already described in Table 7.2)—Cont'd

Galaxy	Description
Fornax	Glow is 20′ in diameter, round with no brightening towards the center (Tom Polakis 13″). Viewed this very difficult dwarf as a subtle brightening of the field, confirmed by tracing around the edge of the halo where the contrast with the background sky was evident (SG 17.5″)
Leo A	Faint, round glow about 4′ across surrounding five 14 mag field stars (Tom Polakis 13″).
Leo I	Pretty easy, elongated E to W 8′ by 6′ with a uniform inner 6′ by 4′, then tapering off gradually. A narrow apparent field eyepiece fares better than a sophisticated wide-field design in eliminating glare from Regulus (Tom Polakis 13″). A few minutes of arc north of Regulus. Large, extremely low surface brightness galaxy, very dim. Fat, not quite possible to tell if elongated – might be round but I think somewhat elongated e/w (I later learned this galaxy was 11′×8′ at PA 80° – not bad!). It is in a star field that 'wraps around' the object on the N, S, and F sides. There is no hint of any condensation whatever; very even brightness (Jeff Medkeff 10″)
Leo II	Appears at low magnification as a brightening of the field measuring 15′ across (Tom Polakis 13″)
Pegasus	Faint, pretty small, elongated 1.8′×1′ in PA 120°, very little brighter in the middle, averted vision makes it somewhat larger, not much. Like most of the dwarf galaxies that I have observed, it is a pretty low surface brightness object (SC 13.1″, 100×)
Pisces	Very faint, pretty small, very little elongated 1.2′×1′ in PA 165°, very little brighter middle at 150×. This is a low surface brightness object (SC 13.1″)
SagDIG	Two dozens of faint stars over a region of 2′ across (Tom Polakis, Larry Mitchell 25″)
Sculptor	Extremely faint, very, very large, little elongated and very little brighter in the middle. After trying to see this huge object with a 16″ scope at low power, I reasoned that we needed a wider field of view and went after it with the little RFT. Every precaution was taken to get fully dark adapted and a cloth was held over the observer's head to block out extraneous light. Using all those precautions, there is a very faint, roundish blob (SC 4.5″ f/4, 16×).Very large, very faint, little elongated 1.5′×1′ in PA 120°, very little brighter middle. Averted vision and the hood do help, but it is still a toughie. This low surface brightness galaxy was held steady with direct vision under these excellent conditions. It grows quite a bit in size and brightness with averted vision. Not easy, but seen (SC 6″)
Ursa Minor	Extremely faint, pretty large, elongated 1.5′×1′ in PA 75°, very, little brighter in the middle. Averted vision is the only way to detect it with a small scope, I never held it with direct vision (SC 6″). Very faint, large, elongated 1.5′×1′ at 60× in a 38 mm eyepiece that gives a one degree field. This is a big, low surface brightness object, so save it for a dark site (SC 13.1″)
WLM	Extremely faint, very large, elongated 2:1 N–S, ~10′×5′. Irregular outline, very low surface brightness. Very weak, if any, central concentration. Best at 100×. This is a very difficult object in less than ideal conditions (SG 17.5″)

Fig. 8.1. NGC 185 in Cassiopeia, a fairly bright companion of M 31 with an interesting dark feature

aperture. On the other hand, observers using smaller richfield instruments under very dark skies have scored a considerable measure of success.

Leo I. A faint dSph discovered by examination of early Palomar Observatory Sky Survey (POSS) plates in 1950 [207]. Its close apparent proximity to the bright star Regulus (α Leo) greatly hindered its discovery. However, if the Regulus is placed outside the field of view, this object is not particularly difficult to observe with moderate aperture.

SagDEG. Discovered only in the mid-1990s, this system is being tidally disrupted by own much more massive Milky Way [207]. Though it has been dispersed over an $8° \times 22°$ ellipse [154] and thus far too diffuse to be seen visually, it does host a small collection of GCs. By far the largest, M 54 (NGC 6715), is second only to Omega Centauri in size and brightness. It has been suggested that it may actually be the core of the galaxy, though this has yet to be proven conclusively.

Sculptor. A faint dSph (Fig. 1.30) that was discovered by Mrs. Lindsay from plates taken by the Bruce telescope in 1938 [207]. With an absolute magnitude of –9.8 mag, this system is only as bright as a gigantic globular like Omega Centauri (NGC 5139), though it has a volume well over one thousand times greater.

Fig. 8.2. NGC 1049, a bright globular cluster in the Fornax dwarf galaxy

Ursa Minor. Even smaller and fainter than the Sculptor dSph, it was not discovered until 1955 via the examination of POSS plates. With an absolute magnitude of –8.8 mag [207], this is one of the smallest and intrinsically faintest galaxies that can be visually detected.

WLM. This is short for its discoverers Wolf (finding it already in 1909), Lundmark and Melotte (Fig. 8.3). A fairly typical dwarf irregular that is perhaps best known for the 16.56 mag globular cluster WLM-1 (see Table 8.11), located about 2′ west of the galaxy's core.

There are other nearby groups, which are spread across large areas of the sky. By using their common distance we can identify their members. A prominent example is the Sculptor group, containing bright galaxies like NGC 55 (Fig. 1.12), NGC 247 (Fig. 7.6), NGC 253 (Fig. 3.1), NGC 300 (Fig. 7.5), and NGC 7793; see Table 7.5 for descriptions.

Hidden Behind the Veil: The IC 342/Maffei Group

Another example of a nearby galaxy cluster is the highly obscured IC 342/Maffei group (Tables 8.3 and 8.4) [145,146]. Due to a distance of only 3.3 Mpc, it is spread widely across the sky. This group is dominated by Maffei I (a giant elliptical), Maffei II (a barred

Fig. 8.3. The "Wolf-Lundmark-Melotte" system (WLM) in Cetus, a very difficult target

spiral), and the large face-on spiral IC 342 [146]. Many of these objects are located deep within the "zone of avoidance" – an area that is heavily obscured by the dust and gas clouds of our Milky Way. The three largest galaxies would be visible to the naked eye if they were not behind this dense veil of dust and gas. Another interesting member is NGC 1569; an example of a Wolf-Rayet galaxy (WR) or starburst galaxy, which has a large population of luminous Wolf-Rayet stars. NGC 1569 contains two extremely luminous super star clusters (see Table 8.11). Much fainter is Dwingeloo 1, a highly obscured large barred spiral discovered by HI radio observations (Fig. 8.4).

Additional Notes

Maffei I. Originally catalogued as a diffuse nebula (Sh2-191), its true character was not recognized until 1968. Obscured by over five magnitudes of extinction, only the innermost 3′ are visible in the telescope [146]. Deep images by Ron Buta and Marshall McCall reveal a far more extensive object. Maffei I is the closest giant elliptical galaxy, with a mass in the neighborhood of several hundred billion suns.

Table 8.3. Members of the IC 342/Maffei group of galaxies

Galaxy	Con	R.A.	Decl	V	V'	$a \times b$	Type	Remarks
Cas 1	Cas	02 06 02.8	+68 59 59	16.0	16.5	1.9×1.6	Im	PGC 100169
Maffei I	Cas	02 36 35.4	+59 39 19	11.4	13.8	3.4×1.7	S0-pec:	UGCA 34
Maffei II	Cas	02 41 55.1	+59 36 15	13.7	16.1	5.8×1.6	SAB(rs)bc:	UGCA 39
Dwingeloo 1	Cas	02 56 51.9	+58 54 42	8.3	8.4	4.2×0.3	SB(s)cd	PGC 100170
UGC 2773	Per	03 32 07.6	+47 47 33	14.1	13.8	1.1×0.8	E6	Double system
IC 342	Cam	03 46 48.4	+68 05 44	8.4	14.9	21.4×20.9	SAB(rs)cd	
UGCA 86	Cam	03 59 58.2	+67 08 20	13.0	14.8	0.8×0.7	Im?	EGB 0427+63
Cam A	Cam	04 25 16.3	+72 48 21	14.5	16.5	3.7×2.1	Im?	EGB 0419+72
NGC 1569	Cam	04 30 49.1	+64 50 43	11.2	13.1	3.7×1.8	IBm	Arp 210, WR galaxy
NGC 1560	Cam	04 32 47.5	+71 52 46	11.4	14.1	9.8×1.5	S(A)sd sp	FGC 71A
Cam B	Cam	04 53 07.1	+67 05 57	15.6	16.0	2.2×1.1	Im?	PGC 166084
UGCA 105	Cam	05 14 15.3	+62 34 48	13.5	16.5	5.5×3.5	Im?	
Mailyan 16	Cam	06 47 45.8	+80 07 26	16.0	15.5	0.7×0.5	S/Irr	PGC 95597

Table 8.4. Observations of galaxies in the IC 342/Maffei group

Galaxy	Description
Cas 1	Very weak brightening, distinct from foreground stars (Frank Richardsen 20″, 630×)
Maffei I	Difficult object visually due to the richness of the surrounding milky way star field. In fact, a group of faint stars is superimposed on the galaxy. Core appears as diffuse haze elongated east–west near the NW edge of a small trapezoid of stars (SG 13″). Very faint, round brightening (Frank Richardsen 20″)
Maffei II	Foreground stars suggest a supposed core (Frank Richardsen 20″)
Dwingeloo 1	Central region seen as weak brightening (Frank Richardsen 20″)
UGC 2773	Fairly faint, small, uniform disk (14″)
IC 342	Faint compact spot (16×70). Compact center; large halo with averted vision (8″, 50×). Pretty faint, very large, round, and much brighter in the middle at 100×. This low surface brightness galaxy was difficult to find from a light polluted site (SC 13.1″). Very unusual galaxy, appears as a very faint, very large glow surrounding a 1′ high surface brightness core which increases to a bright stellar nucleus. Irregular halo is difficult to trace but ~10′ diameter and has a number of superimposed stars including a striking 6′ string of six 10.5–12 mag star oriented NW–SE on the SW side of the halo. The core forms a small triangle with two similar superimposed 11 mag stars 1.0′ N and 2.0′ NE (SG 17.5″). Extremely large, diffuse object with a weak, faintly defined spiral structure over 20′ across. Would be an extremely impressive object if it was not dimmed by nearly 2 mags (RJ 24″)
UGCA 86	Very faint, small, round (20″)
Cam A	Extremely weak brightening, difficult to separate in crowded field (Frank Richardsen 20″)
NGC 1569	Fairly bright, small, elongated. Located just S of a 9 mag star (SG 8″). Very bright, elongated 2:1 WNW–ESE, high surface brightness, elongated bright core, mottling suspected (SG 13″)
NGC 1560	Very faint, fairly large, edge-on SSW–NNE, low even surface brightness. Appears as a ghostly streak (SG 8″). Fairly faint, very large, 6′×1′, low surface brightness edge-on SSW–NNE. Broad weak concentration with no distinct core but there a central 2′ brightening. A 13 mag star is embedded on the preceding side of the NNE extension. The galaxy appears to extend very faintly beyond this toward a 12 mag star further N. Another 13 mag star is superimposed at the SSW end and a brighter 11.5 mag star is just following the tip of this extension (SG 17.5″)
UGCA 105	Faint, but clearly visible, slightly structured center, elongated halo (Frank Richardsen 20″)

Maffei II. Like its neighbor (Maffei I), this was originally misclassified as a diffuse nebula (Sh2-197) and discovered, as a galaxy, by Paolo Maffei in 1968 [146]. Nearly 6 mag of galactic extinction have reduced the "visible part" to only a diffuse glow a mere 1′ across. But deep exposures and extensive image processing produce a different picture – a huge barred spiral galaxy measuring nearly 20′ across! If it were not for this severe extinction, Maffei I and II would be the most spectacular galaxy pair in the northern heavens. They would appear as a pair of 6 mag objects, separated by only one degree – and far more impressive than M 81/M 82!

Fig. 8.4. Dwingeloo 1 in Cassiopeia, a heavily obscured member of the IC 342/Maffei group

Dwingeloo 1. Another large barred spiral galaxy obscured by at least 5 mag of extinction (Fig. 8.4). This object would shine at 8 mag and would likely have been classified as a "Messier Object" if placed in a better part of the sky.

IC 342. A large spiral inclined about 25° to our line of sight. Similar to size and luminosity to our own Milky Way, it is nearly 2/3 the apparent size of the Full Moon [146]. Though dimmed by an extinction of 2 mag, it is nonetheless an impressive object in a large telescope.

NGC 1569. An unusual irregular galaxy undergoing a major episode of star formation. One of the closest known "starburst galaxies," it has a very high surface brightness and an irregular, complex structure in large instruments. See also Tables 8.11 and 8.12, and the following "Additional Notes."

UGCA 86. A small satellite galaxy of IC 342 located about 1.5° SE of the main system.

The Nine Holmberg Dwarfs

Having already met a number of nearby dwarfs in our tour of the Local Group, let's not forget to mention nine classic objects first described by Eric Holmberg (Table 8.5). All,

Table 8.5. The nine Holmberg dwarfs

Holmberg	Con	U2	R.A.	Decl	V	V'	$a \times b$	PA	Type	Remark
I	UMa	14	09 40 29.1	+71 11 01	12.6	15.3	4.0×3.3	117	IAB(s)m	UGC 5139
II	UMa	14	08 19 03.9	+70 42 55	10.7	14.8	8.5×6.0	15	Im	UGC 4305
III	Cam	6	09 14 48.3	+74 13 56	12.4	14.2	2.3×2.1	150	SAB(s)c	UGC 4841
IV	UMa	23	13 54 44.4	+53 53 58	13.4	15.1	4.6×1.2	18	IB(s)m sp	UGC 8837
V	UMa	23	13 40 39.9	+54 19 55	12.7	13.9	2.5×1.5	110	SAB(rs)c	UGC 8658
VI	Eri	156	03 24 48.4	−21 20 11	12.6	14.0	2.1×1.9	144	SAB(rs)d	NGC 1325A
VII	Vir	91	12 34 45.3	+06 17 55	14.0	13.8	1.2×0.8	153	Im	UGC 7739
VIII	CVn	53	13 13 17.6	+36 12 50	13.1	14.7	2.2×2.2		IAB(s)m	UGC 8303
IX	UMa	14	09 57 34.5	+69 02 42	14.1	16.1	2.8×2.5	135	Im	UGC 5336

except Holmberg III, V, and VI are of the "magellanic" type of irregular galaxy in de Vaucouleurs' classification. Holmberg I, II (Fig. 8.5), and IX belong to the M 81 group of galaxies, located at a distance of 3.8 Mpc [147]; Holmberg IX is only 9′ W of M 81. Holmberg III is a possible member of the IC 342/Maffei group (Table 8.3). Holmberg IV and V are members of the M 101 group (8 Mpc). The rest, Holmberg VI (18 Mpc), Holmberg VII (15 Mpc?), and Holmberg VIII (11 Mpc) are probably mere field galaxies. All are low surface brightness objects, though some are visible in small telescopes (Table 8.6) [148].

Non-stellar Objects in Nearby Galaxies

Some nearby galaxies can offer an interesting variety of nonstellar objects. Classic examples are the nebulae and clusters in the Magellanic Clouds [126], like the giant HII region 30 Doradus [149]. Another favorite target is M 31, offering all classes of objects (Table 8.7), including numerous globular clusters. The most celebrated is G 1, a gigantic cluster that is even larger and more massive than our own Omega Centauri. Resolved into a blaze of stars by the HST, it has an ellipsoidal shape and shows rotation – which is highly unusual for a globular. Like Omega Centauri, it may have once been the core of a dwarf elliptical (Fig. 8.6). Many keen observers have studied nonstellar objects in M 31 and its

Fig. 8.5. The dwarf galaxy Holmberg II in Ursa Major

Table 8.6. Observations of Holmberg dwarfs with various apertures

Holmberg	Description
I	Very faint dwarf, visible only at 115× with averted vision. It appeared less than 1′ in diameter and round (SG 17.5″). Small round patch with stellar core (Frank Richardsen 20″)
II	Difficult (Brian Skiff 6″ refractor). Round, about 5′ across, and near an equilateral triangle of 13 and 14 mag stars. I was unable to pick out any of the HII knots (SG 17.5″)
III	Difficult, diffuse patch (14″).
IV	Difficult, faint, oval patch (14″). Elongated, slightly brighter middle (Frank Richardsen 20″)
V	Very difficult, only nucleus (8″). Bright nucleus, spiral arms difficult (Frank Richardsen 20″)
VI	Difficult, round, brighter middle (20″)
VII	Faint patch (Frank Richardsen 20″)
VIII	Difficult (14″). Irregular patch (Frank Richardsen 20″)
IX	Very difficult, faint diffuse patch (Frank Richardsen 20″)

Table 8.7. Nonstellar objects in M 31 and M 110 (SC = star cloud, OC = open cluster, GC = globular cluster)

Object	Type	R.A.	Decl	V	Size	Remarks
NGC 206	SC	00 40 32.3	+40 44 18	15.0	4.2′	A 78 (Association)
C 202/203	OC	00 42 05.6	+40 57 09	14.5	10″/ 12″	Double cluster (sep. 15″)
C 306	OC	00 44 56.2	+41 31 27	15.0	20″	
C 312	OC	00 45 10.5	+41 36 57	15.5	25″	
G 1	GC	00 32 46.5	+39 34 41	13.7	30 ″	Mayall II, core of dwarf elliptical?
G 63	GC	00 40 32.4	+41 39 19	15.6	6″	in M 110 (Hubble 2)
G 73	GC	00 40 32.3	+40 44 18	14.6	6″	in M 110 (Hubble 3)
G 76	GC	00 40 59.1	+40 35 48	14.3	12″	
G 78	GC	00 41 01.3	+41 13 45	14.3	9″	
G 213	GC	00 43 12.6	+41 07 21	14.7	8″	
G 219	GC	00 43 18.0	+39 49 13	15.1	9″	Mayall IV, Mrk 959
G 272	GC	00 44 14.8	+41 19 08	14.8	9″	
G 280	GC	00 44 28.5	+41 21 30	14.2	9″	

companions, with a wide range of apertures (Table 8.8) [150]. For example, Art Russell (Atlanta, USA) has seen over 50 of these objects with an 18 in. telescope. Their "bible" is still Paul W. Hodge's *Atlas of the Andromeda Galaxy* [151]. Further information can be found in [152,242], in Luginbuhl & Skiff, and for the extensive globular cluster system, in [153,244].

Based on recent research, some of the nearest extragalactic globulars could be M 79, NGC 1851, NGC 2298, and NGC 2808, associated with the newly discovered dwarf ellip-

Fig. 8.6. G1 near M 31: globular cluster or core of dwarf elliptical galaxy?

tical galaxy in Canis Major. A similar case could be made for M 54, Terzan 7, Terzan 8, and Arp 2, which appear to belong to the nearby dwarf in Sagittarius (SagDEG) [154]. While M 54 and M 79 can be seen in binoculars and the three NGC globulars are visible in a small telescope, the Terzan and Arp objects are quite challenging.

A few of the very brightest single supergiant stars in M 31 are visible in a large amateur telescope. Many of these stars are in the super-association complex NGC 206. The brightest is AF And, varying between 13.3 and 17.5 mag. It is an example of a "luminous blue variable" (LBV) and can be readily seen (at maximum) with a 13 in. telescope. An even brighter example (concerning its absolute magnitude) is V1 in the galaxy NGC 2366, glowing dimly at 17.4 mag from a distance of 2.4 Mpc. Brilliant hypergiant stars like these are among the most luminous single objects in the entire sky.

Besides M 31, the Triangulum Nebula M 33 is a worthwhile target (see [155,156] or Luginbuhl & Skiff). It has a large collection of NGC/IC objects, which are associated with bright HII regions or star clouds in the loose spiral arms. Most prominent is NGC 604 (Fig. 7.7), an HII region that is far brighter and larger than our own Orion Nebula complex. It is similar in size to that of the 30 Doradus (Tarantula) in the LMC, and some of the irregular, filamentary details are visible in larger instruments. As the brightest supergiants at magnitude 16–16.5, some observers using large (over 20 in.) telescopes have reported resolving some of the stars in this object. Ron Buta using a 36 in. once remarked that *"NGC 604 appears like a nebulous version of the Pleiades as seen by the naked eye."* Selection of other interesting objects in M 33 are listed in Table 8.9 [230].

Table 8.8. Visual descriptions of nonstellar objects in M 31 and M 110

Object	Description
NGC 206	Very faint, low contrast, oval (14″). Fairly faint, fairly large, elongated 5:2 north–south, 4.0′ × 1.6′, low and uneven surface brightness, a few very faint stars are just visible over the surface including a brighter star at the south tip, located over 1° southwest of the core of M 31 (SG 17.5″)
C 202 C 203	C 202 faint, extremely small, ~6″, easily visible as the slightly fainter southern member of a very close nebulous pair with C 203 15″ north, just cleanly resolved at 220×. C 203 faint, extremely small but nonstellar, similar size but slightly brighter than C 202. Located southwest of core of M 31 and 9′ northwest of M 32 (SG 17.5″)
C 306	Fairly faint, appears as a 13 mag star with a 12 mag star located 2′ northwest, easily visible at 100× and prominent at 220×. Located along eastern flank north of core (SG 17.5″)
C 312	Threshold object, very small elongated glow requiring averted, located 2′ north of line connecting two stars 12–13 mag oriented WNW/ESE with a separation 6′. I probably glimpsed the unresolved glow of C 311, C 312, C 313 with a total length of 1.5′ (SG 17.5″)
G 1	Bright spot, almost stellar, in triangle with two stars (14″). Easily picked up sweeping at 220× because on first glance it appears to be a close but cleanly resolved 14 mag triple star! On closer viewing, the central "star" is soft or slightly nebulous and surrounded by a very faint and small halo <10 in. diameter. The foreground companion stars are situated at SW and NW edges of the halo (SG 17.5″)
G 63	Visible with 16″ at 400×.
G 73	Very faint, visible with direct vision as a 15 mag star, on line with a 8.5 mag star 2.5′ north and 10 mag star 6.5′ north. Located 6′ east of the center of M 110 (SG 17.5″)
G 76	Faint, easily visible as a stellar object of 14.2 mag located in small "W" asterism and very close northwest of a similar star 14.2–14.4 mag, three similar brighter stars within this "W" also form a distinctive obtuse triangle. Located on south side of southwest tip of M 31 (SG 17.5″)
G 78	Faint, easily visible as a star of 14.3 mag, forms west vertex of slightly obtuse triangle with two stars of 12.5 mag oriented north–south and situated 2′ east. Located roughly 30′ west of the core of M 31 (SG 17.5″)
G 213	Very faint, visible continuously with direct vision as a 15 mag "star," an 11.5 mag star is 1.5′ east–northeast. Located ~10′ southeast of the center of M 31 near the edge of the central haze (SG 17.5″)
G 219	Picked up at 220× as a very faint 15 mag "star," possibly quasistellar. At 380×, appears as barely nonstellar glow, ~2″ in diameter. Easily visible at this power (SG 17.5″)
G 272	Very faint, appears as a 15 mag "star," just visible continuously with direct vision, a 10.5 mag star is 1.5′ northwest (SG 17.5″)
G 280	Faint, appears as an easily visible 14.3 mag star, perhaps slightly fainter than G 76 and G 78, a similar star is 1′ east–southeast, also G 272 is in field 4′ southwest. Located along east side of northeast extension of M 31 outside of prominent central region (SG 17.5″)

Object	Type	R.A.	Decl	V	Size	Remarks
NGC 588	HII + SC	01 32 45.7	+30 38 56	13.5	40	
NGC 592	HII + SC	01 33 12.0	+30 38 47	13.0	42	
NGC 595	HII	01 33 34.0	+30 41 32	13.5	30	
NGC 604	HII	01 34 32.6	+30 37 04	12.0	120	
IC 131	HII + SC	01 33 11.1	+30 45 10	14.0	20	
IC 132	HII + SC	01 33 16.0	+30 56 42	13.5	42	
IC 133	HII + OC	01 33 15.8	+30 53 05	14.0	24	
IC 135	HII	01 34 15.5	+30 37 10	14.0	25	
IC 136	OC	01 34 13.5	+30 33 40	14.5	18	
IC 137	OC	01 33 38.8	+30 31 23	14.0	36	
IC 139	OC	01 33 59.3	+30 34 33	14.0	30	
IC 140	OC	01 33 58.1	+30 33 02	13.0	30	
IC 142	SC	01 33 55.8	+30 45 22	14.2	30	
IC 143	HII	01 34 11.1	+30 46 41	14.0	30	
C 27	GC	01 34 43.8	+30 47 37	16.5	5	
C 39	GC	01 34 49.7	+30 21 55	15.9	8	Mayall C

Table 8.9. NGC/IC objects (SC = star cloud) and the brightest globulars in M 33 (size in ″)

Nonstellar objects suitable for visual observation can also be found in other galaxies of the Local Group, and even in galaxies farther beyond, such as M 101 [112]. Table 8.11 show prominent examples. Some are even more conspicuous than their hosts, like in the case of the globular cluster NGC 1049 of the Fornax dwarf spheroidal. For other globular clusters beyond the local group you may wish to check the catalogue of Harris [157].

Additional Notes

NGC 1049. Excluding the GCs that may be associated with the SagDEG and Canis Minor systems; this is the largest and brightest extragalactic globular cluster available to the amateur (Fig. 8.2). Four other GCs are visible in the Fornax System – Fornax 2, 4, and 5 are around 13.5 mag, while Fornax 1 is much smaller and fainter at 15.6 mag [154].

NGC 1569. This starburst galaxy (a member of the nearby IC 342/Maffei galaxy group, see Table 8.5) is the home of a number of brilliant blue SSCs, of which two designated A and B are visible in larger scopes. Both form a tiny "double star" of 6″ separation. However, in reality these are luminous "proto-globulars" nearly 100× the brightness of Omega Centauri [154]. At Omega's distance, they would be around –1 mag and larger than the Full Moon! Other good examples can be found in M 82, NGC 1705, NGC 4449, and NGC 6946.

NGC 2363. This is a massive star forming complex in NGC 2366 and is at least as large as 30 Doradus in the LMC or NGC 604 in M 33 (Fig. 7.7). Intensely studied by the HST, this complicated region hosts a hypergiant star or luminous blue variable (LBV) that rivals our Milky Way's own Eta Carinae. Varying from 17.8 to 21.5 mag, it is in range of amateur CCDs and the largest (meter class) telescopes.

Table 8.10. Visual descriptions of M 33 objects

Object	Description
NGC 588	Extremely faint nebulosity requires averted vision. Located 14′ W of the center of M 33 and forms the W vertex of a very obtuse isosceles triangle with NGC 592 6′ W and NGC 595 (SG 17.5″)
NGC 592	Faint nebulosity 9′ W of the core; see NGC 588 (SG 17.5″)
NGC 595	Very faint nebulosity 4′ NW of the center. Situated just off the W edge of the beginning of the spiral arm which extends N from the core on the W side; see NGC 588 (SG 17.5″)
NGC 604	Fairly bright, round (SG 8″). Bright HII region located 12′ NE of the core. Situated at the end of the large spiral arm of M 33, which extends N and then E of the core. Bright, small, round (SG 13″)
IC 131	Very faint, very small, round, 10″ diameter. This HII region is located 10′ NW of the center of M 33 near a wide pair (50 in.) of 11/12 mag stars. A 14 mag star is close by and at first I thought this was IC 131 (SG 17.5″)
IC 132	Faint but easily visible HII knot of 20″ diameter. Located 1′ N of a pair of 13 mag stars at 10 in. separation and 1.6′ W of a 9 mag star (SG 17.5″)
IC 133	Faint, diffuse, hazy HII region of 35″ diameter at the NW end of M 33 15′ NW of the center. Forms a "pair" with IC 132 3.4′ N. This object is larger than IC 132 at times with averted vision but has a lower surface brightness (SG 17.5″)
IC 135	Fairly faint, fairly small, 1′ diameter. This HII region is located 6′ ESE of the center of M 33 (SG 17.5″)
IC 136	Very faint, ill-defined hazy region between IC 135 3.5′ N and an 11.5 mag star 2.5′ SSE (just W of the line connecting these objects). Appears as a slightly locally brighter region of 30″ diameter and not as noticeable as the other IC HII regions – would have passed over if casually sweeping galaxy (SG 17.5″)
IC 137	Very faint HII knot in M 33 located at the S end of a spiral arm 10′ SSW of center (SG 17.5″)
IC 139	Fairly prominent elongated HII region just following a 13 mag star 5.4′ SSE of the center of M 33. Extended ~N–S, perhaps 2.0′×0.5′ and consists of two brighter knots at both ends (SG 17.5″)
IC 140	Located SW of IC 139. Easy knot, ~1′ in diameter with ill-defined edges. There is a second knot close W which is slightly fainter (SG 17.5″)
IC 142	Fairly faint, very small, round. Stands out nicely 6′ N of the center of M 33. Either contains a stellar spot near the center or a faint star is superimposed (SG 17.5″)
IC 143	Very faint HII region in M 33 located 8′ NNE of the center and 5′ W of NGC 604. Appears very faint, small, round, 20″ diameter. There is a 13.5 mag star 2′ SE and close WNW of this star is also a faint, hazy patch of nebulosity (SG 17.5″)
C 27	Located about 2.5′ east of NGC 604, this cluster was very faint and difficult to observe. It was distinctly stellar, with no evidence of an extended envelope visible (RJ 20″, 260×)
C 39	This is the brightest globular cluster, located ~22 minutes southeast of M 33's nucleus. It was visible with direct vision as a slightly fuzzy "star" of 16 mag. The extended halo or envelope was <2″ in diameter. It was the only globular I observed that had a nonstellar appearance (RJ 20″, 260×)

Table 8.11. NGC/IC objects in other galaxies (size in ″)

Galaxy	Con	Object	Type	R.A.	Decl	V	Size	Remarks
M 101	UMa	NGC 5447	HII+SC	14 02 29.0	+54 16 21	13.5	60	
		NGC 5449	HII+SC	14 02 28.2	+54 19 53	14.0	60	
		NGC 5450	HII	14 02 28.9	+54 16 22	13.0	60	
		NGC 5451	SC	14 02 36.5	+54 21 49	14.0	20	
		NGC 5453	SC	14 02 56.7	+54 18 31	13.8	30	
		NGC 5455	SC	14 03 01.0	+54 14 27	13.0	25	Compact galaxy?
		NGC 5458	HII	14 03 12.4	+54 17 56	14.0	36	
		NGC 5461	SC	14 03 41.5	+54 19 05	14.0	40	
		NGC 5462	SC	14 03 53.0	+54 22 02	13.5	60	
NGC 55	Scl	IC 1537	SC	00 15 49.5	−39 15 42	15.0	30	ESO 249-1
NGC 1569	Cam	A	SSC	04 30 48.2	+64 50 59	14.8	6	Double cluster
		B	SSC	04 30 49.0	+64 50 53	15.6	6	
NGC 1705	PIC	A	SSC	04 54 13.2	−53 21 36	14.3	3	
NGC 2366	Cam	NGC 2363	HII+SSC	07 28 29.7	+69 11 34	13.0	90	Massive star forming complex
NGC 2403	Cam	NGC 2404	SSC	07 37 07.0	+65 36 40	14.5	25	
NGC 2848	Hya	NGC 2847	SC	09 20 08.6	−16 31 02	15.0	10	
NGC 2874	Leo	NGC 2875	SC	09 25 48.7	+11 25 56	15.5	10	
NGC 2903	Leo	NGC 2905	SC	09 32 11.8	+21 31 07	15.0	10	
NGC 3184	UMa	NGC 3180	HII	10 18 10.7	+41 26 57	15.0	15	
		NGC 3181	SC	10 18 11.5	+41 24 48	14.8	20	
NGC 4214	CVn	HK 2	HII	12 15 41.0	+36 19 11	14.5	20	
NGC 4395	CVn	NGC 4399	HII	12 25 42.9	+33 31 00	14.0	60	
		NGC 4400	HII	12 25 55.9	+33 30 57	14.5	25	
		NGC 4401	HII	12 25 57.9	+33 31 38	14.0	40	
NGC 4449	CVn	1	OC	12 28 11.0	+44 05 37	15.5	10	
NGC 4559	Com	IC 3550	HII	12 35 51.8	+27 55 57	14.5	25	
		IC 3551	HII+SC	12 35 53.8	+27 57 48	14.5	20	
		IC 3552	SC	12 35 56.2	+27 59 28	15.5	10	

		IC 3555	SC	12 35 56.0	+27 59 24	15.0	30	
		IC 3563	SC	12 35 07.1	+27 55 36	15.2	30	
NGC 4818	CVn	IC 3668	SC	12 41 32.9	+41 07 26	14.5	30	
	CVn	IC 3669	SC	12 41 37.5	+41 08 25	15.5	15	
NGC 4654	Vir	IC 3708	SC	12 43 51.8	+13 08 12	15.0	15	
NGC 5907	Dra	NGC 5906	SC	15 15 52.1	+56 19 45	14.5	10	
NGC 6822	Sgr	IC 1308	HII	19 45 05.2	–14 43 17	14.0	36	Hubble X
		Hubble V	HII	19 44 52.4	–14 43 11	14.5	30	
NGC 6946	Cep	1447	SSC	20 34 31.5	+60 08 15	15.7	30	
WLM	Cet	WLM-1	GC	00 01 49.5	–15 27 31	16.1	8	
Fornax Dwarf	For	NGC 1049	GC	02 39 48.2	–34 15 29	12.9	40	ESO 356-3

NGC 4395. This is a large, diffused, weakly developed barred spiral of low surface brightness. Over a half dozen large HII regions may be glimpsed by the observer, of which three have NGC designations (NGC 4399, NGC 4400 and NGC 4401) [205]. This galaxy also has the distinction of being one of the faintest known of all Seyfert galaxies.

M 101. A giant Sc galaxy (Fig. 4.4) that is home to hundreds of HII regions/OB associations, of which nine have NGC designations (see Table 8.11) [225]. The brightest, NGC 5471, is one of the largest known: a "hypergiant" HII region according to Waller & Hodge. Its size and mass is comparable to the small starburst galaxy NGC 1569.

Moving farther out, nonstellar objects in galaxies eventually become too small and faint to be visible. But there are certain stellar objects that are easily observed: supernovae! Such events are rare for a specific galaxy (like the Milky Way), but are frequent in a large sample of galaxies (Fig. 8.7). Many amateurs and amateur search teams are scanning the heavens visually or with CCDs in a systematic manner [158]. Some teams such as Tim Puckett's have made an impressive number of discoveries (over 100) in the past decade [228,240]. A recent, spectacular event was the supernova in M 51 (SN 2005cs), discovered by the German amateur Wolfgang Kloehr [238].

Great Distance: AGN, Quasars, and BL Lacertae Objects

We will now completely jump over the mid-range (as there are enough examples given), to the extremely remote, super luminous cosmic objects: infrared galaxies (visible through starburst phenomena), "active galactic nuclei" (AGN), quasars and BL Lacertae objects. The following Table 8.13 contains a selection of 20 prominent examples found in the northern sky (for a similar list compare [159]). The BL Lac Object 3C 66A, surrounded by three UGC galaxies (see Tables 6.1 and 6.2), was already presented as an example of a starhopping tour.

Many objects are easily visible in an 8″ telescope, but it is difficult to say which is the brightest. Among the favorites are 3C 273, 3C 465, KUV 1821+643, S5 0716+71, and PGC 61965 (Fig. 8.8). Due to variability a few objects can even get brighter than 12.5 mag, e.g., W Com or Mrk 421, are visible with 4″ aperture. The most prominent and best known quasar, 3C 273, is slightly variable too. The first one discovered, 3C 48, is normally quite faint, but can be seen with a 12″ when at maximum brightness. Some objects show bursts, in which the brightness can increase by several magnitudes in a span of a few days.

Mrk 421 and Mrk 501 are among the nearest, followed by 3C 465; while the most remote example is PG 1634+706. This distant object also has the highest absolute magnitude in the list. At an astounding −30.3 mag, it is over 4,000 times brighter than the Andromeda Galaxy (M 31). Most objects are stellar (at all magnifications), but some appear a bit fuzzy by showing a stellar core, surrounded by a small halo.

This collection lists only 1/4 of all QSOs visible with 10–12″. With a 14″ you might reach around 250. Each year many new objects are discovered, and some of them are quite bright [92].

We have already mentioned that for an observation it is important to have a good finding chart, which shows stars down to 15 mag (the GSC is a sufficient source). Most

Table 8.12. Visual descriptions of nonstellar objects in other galaxies (from Table 8.11; sorted by NGC/IC number)

Object	Description
NGC 1049	Pretty faint, small, round, much brighter in the middle, averted vision makes it grow at 135×. A little grainy at 165× (SC 13.1″).
NGC 1569 A, B	They form a bluish "double star" with PA 110°, and ~6″ separation (RJ 20″). Both objects observed with 20″ (Jens Bohle)
NGC 1705 A	Faint 1.3′×1′ haze, elongated NE–SW, which rises smoothly to the middle and shows the SSC NGC 1705-A as a bright 14 mag stellar spot W of centre. At 650× there is some subtle mottling in the core but no specific detail can be seen. A 14 mag star is 1′ NW and a number of 12–14 mag stars are in the surrounding field (Michael Kerr 25″)
NGC 2363	A massive H II region/star forming complex that has a higher surface brightness than the host galaxy (NGC 2366). Visible as an oval diffuse spot in medium scope, it takes on a more irregular appearance in large aperture (RJ, 13″, 20″)
NGC 2404	Extremely small emission "knot" at the E end of NGC 2403 (SG 13″) Located at the end of the northern spiral arm of NGC 2403. Appears fairly faint, small, round, clearly nonstellar(SG 17.5″)
NGC 2905	Very faint knot or arc at NE edge of arm of NGC 2903 (SG 13″) Very large knot or arc at the NNE edge of a spiral arm in NGC 2903. Easily visible with averted vision (SG 17.5″)
NGC 4214 IA, IB	Both associations observed with 20″ (Jens Bohle)
NGC 4399	Faintest of three HII knots observed in NGC 4395. Appeared extremely faint and small, 10–15 in. in size and situated 2.3′ SW of the ill-defined core on a line with a 14.5 mag star to the NE of the core. Required averted vision to confirm (SG 17.5″)
NGC 4400	Very small HII knot in NGC4395 situated 0.9′ SSW of brighter NGC 4401. Shows up well at 220×, although only 15 in. in size and no other details (SG 17.5″)
NGC 4401	Brightest HII region in NGC 4395 located ~2′ SE of the ill-defined core. Fairly easy at 220× (the galaxy looses its identity at this power!), as an irregular 25 in. knot. Off the south side is a second fainter knot (NGC 4400) (SG 17.5″)
NGC 4449 1	Observed with 20″ (Jens Bohle)
NGC 5447	This is a knot in an outer arm of M 101 on the western side. Easily visible, compact, round. Located symmetrically opposite from NGC 5462 on the opposite side of the core (SG 13″) Brightest HII region on the preceding side of M 101 located 7.8′ SW of center. Appears as a very elongated glow NW–SE situated just S of a 13.5 mag star. A very small knot is partially resolved at the N edge within a common halo with the extension to the SE (SG 17.5″)
NGC 5449	Extremely low contrast HII knot in M 101. Highly suspected hazy spot 3.5′ N of NGC 5457 but difficult to confirm (SG 17.5″)
NGC 5450	This is the bright HII region on the W side of M 101 8′ SW of center. Connected with NGC 5447. Appears as a very elongated glow NW–SE just S of a 13.5 mag star. A very small knot is partially resolved at the N edge (NGC 5447) within a common halo with NGC 5450 (SG 17.5″)
NGC 5451	This is a difficult, low contrast HII region in M 101 located ~5′ WNW of center. Appears very faint, extremely small, round, starry center? (SG 17.5″)

(*Continued*)

Table 8.12. Visual descriptions of nonstellar objects in other galaxies (from Table 8.11; sorted by NGC/IC number)—Cont'd

Object	Description
NGC 5453	This low surface brightness HII region in M 101 was barely distinguishable at 220× as a very low surface brightness enhancement superimposed on the background glow of a spiral arm 3.4′ SW of center (SG 17.5″)
NGC 5455	Fairly faint HII region in M 101 located 6.6′ SSW of center. Very small, round, 15 in. diameter. Appears a compact but nonstellar knot forming an isosceles triangle with two 13 mag stars 2.3′ NE and 2.3′ NW (SG 17.5″)
NGC 5458	Knot in M 101 located just S of the core. Appears as a barely nonstellar spot (SG 13″) Low contrast 25 in. knot superimposed on the main body of M 101 3′ due S of center. Visibility is hindered as superimposed on the brighter background of the central region (SG 17.5″)
NGC 5461	This is a knot in M 101 located in the spiral arm which trails to the E. Appears as a very diffuse, fairly small knot (SG 13″) Fairly faint knot in the trailing arm of M 101 4.5′ SE of center. Appears slightly elongated, ~25 in. × 15 in., fairly high surface brightness. Contains a very small brighter center or a star is superimposed (SG 17.5″)
NGC 5462	Knot in M 101, located in the same arm as NGC 5461 but further to the E (SG 13″) This is an easily visible, compact, round knot on the opposite side of the core as NGC 5447. Moderately bright elongated knot in M 101, extended 3:1 SW–NE, ~50 in. × 20 in. One of the largest and brightest HII regions in M 101 (SG 17.5″)
NGC 6946 1447	Clearly visible in 20″ (Jens Bohle)
IC 1308 Hubble V	HII region on the NE edge of NGC 6822. At 82× and OIII filter appears as a faint, very small but clearly nebulous round knot. Estimate 14 mag. A 12 mag star lies 2′ SE (very close double on the POSS). Forms a pair with similar Hubble V 3′ W (SG 17.5″)

catalogues of quasars are not designed for amateur use. The first usable collection was the *Handbook of Quasi-stellar and BL Lacertae Objects* by Eric Craine [160]. It contains 186 objects brighter than 17 mag with identifications and finding charts. A source, designed for amateurs is the *Catalogue of Bright Quasars and BL Lacertae Objects* (1984) [161], listing 222 objects with $\delta > -20°$ and brighter than 16.5 mag.

There is an exclusive club of variable galaxies appearing stellar (AGN, quasars, BL Lac objects), which were once catalogued as "variable stars." The prototype is BL Lac, varying between 13 and 15.5 mag. The following Table 8.14 (based on [96]) shows the brightest cases (BL Lac and W Com are already listed in Table 8.13). V362 Vul is a stellar appearing starburst galaxy, heavily obscured by the Milky Way (Fig. 8.9). Observations are collected in Table 8.15.

Much more difficult to observe are gravitational lensed quasars and quasar pairs, because there are so few "bright" examples (Tables 8.16 and 8.17). As both tables show, we're now reaching extreme cosmological distances. A few lensed quasars can be observed in a 14″: the "Double Quasar" (Fig. 8.10), the "Triple Quasar", HE 1104–1805, and UM 425. With excellent seeing, the double quasar can be resolved with an

Fig. 8.7. Supernova 1998s in NGC 3877 (Ursa Major)

aperture of around 20″. APM08279+5255, a gravitational lensed quasar in Lynx, might be the most distant object visible in an amateur telescope of the 20″ class. The brightest real double quasar, HS 1216+5032, can be observed with a 20″, but without any chance of separating the components. The pairs CT 344 and PHL 1222 should be visible in a 20″ too, while UM 425 shows a promising separation but the second image is an extremely faint 20.5 mag object. Thus in most cases you will see only one stellar object – if at all!

Additional Notes

Q 0957+561. The famous double quasar in Ursa Major (Fig. 8.10). Its located near a "y"-shaped asterism about 14′ NNW of the bright field galaxy NGC 3079. It large telescopes it appears as a tiny "double star", with components of 16.7 and 17.0 mag, though this is somewhat variable. The lensed components are separated by only 6″ (PA ~ 80°), so use high power (over 300×) for the best results. Though some observers have resolved it with moderate sized telescopes under very dark and stable conditions – the best results are with instruments 15″ or greater in size. While in the neighborhood, check out the galaxy. It's a pretty edge-on measuring 8′×1.4′ (11.5 mag) and features a somewhat warped disk.

Q 2237+0305. Best known as "Einstein's Cross" (Fig. 2.7), this is perhaps the most famous lensed quasar in the sky. The four lensed components are extremely faint (17.4, 17.4, 18.4, and 18.7 mag) and form a square or cross-like pattern. The separation ranges

Table 8.13. Twenty bright northern Quasars (Q) and BL Lacertae objects (BL); Dist = distance in Mpc; appearance is for a 12″ under ordinary seeing (*d/a* = direct/averted vision); *v* = variable

Object	Type	Con	R.A.	Decl	*V*	M_B	*z*	Dist	Appearance
I Zw 1	Q	Psc	00 53 34.9	+12 41 36	14.0	–23.4	0.06	250	Stellar (d)
3C 48	Q	Tri	01 37 41.3	+33 09 35	16.2 v	–25.2	0.37	1343	Stellar (d)
3C 66A	BL	And	02 22 39.6	+43 02 08	15.0 v	–26.5	0.44	1588	Stellar (d)
VII Zw 118	Q	Cam	07 07 13.2	+64 35 59	14.6	–23.1	0.08	322	Stellar with halo (a)
S5 0716+71	BL	Cam	07 21 53.4	+71 20 36	13.8 v				Stellar (d)
OJ 287	BL	Cnc	08 54 48.8	+20 06 30	15.4 v	–25.5	0.31	1141	Stellar (a)
Mrk 421	BL	UMa	11 04 27.2	+38 12 32	12.9 v	–22.9	0.03	129	Stellar (d)
4C 29.45	Q	UMa	11 59 31.9	+29 14 45	14.4 v	–28.6	0.73	2431	Stellar (a)
W Com	BL	Com	12 21 31.7	+28 13 58	15.0 v	–22.8	0.10	411	Stellar (d)
3C 273	Q	Vir	12 29 06.7	+02 03 08	12.8 v	–26.9	0.16	622	Stellar (a)
3C 305	Q	Dra	14 49 21.6	+63 16 14	13.7	–23.3	0.04	174	Stellar with halo (a)
PG 1553+113	BL	Ser	15 55 43.1	+11 11 24	15.0	–26.8	0.36	1323	stellar (d)
PG 1634+706	Q	UMi	16 34 29.0	+70 31 33	13.5	–30.3	1.34	4033	stellar (a)
Mrk 501	BL	Her	16 53 52.2	+39 45 36	13.7	–22.4	0.03	137	stellar with halo (d)
I Zw 187	BL	Her	17 28 18.6	+50 13 11	15.3 v	–21.1	0.06	226	Stellar (a)
KUV 1821+643	Q	Dra	18 21 57.2	+64 20 36	13.3	–27.1	0.30	1111	Stellar (d)
PGC 61965	Q	Dra	18 30 23.1	+73 13 10	13.9	–23.9	0.12	491	Stellar (d)
BL Lac	BL	Lac	22 02 43.3	+42 16 39	14.7 v	–22.4	0.07	282	stellar (d)
MR 2251–178	Q	Aqr	22 54 05.9	–17 34 55	14.3	–23.1	0.07	278	stellar (a)
3C 465	Q	Peg	23 38 29.4	+27 01 52	13.3	–23.1	0.03	129	Nonstellar (a)

Fig. 8.8. The bright quasar PGC 61965 in Draco

from 1.6″ to 1.8″, and when combined with the exceedingly faint magnitudes it make this one of the most challenging objects in the sky [196]. The "cross" is produced by a small 15 mag foreground galaxy CGCC 378–15. This tiny elongated galaxy measures a mere 1.1′×0.5′, and is faintly visible in larger scopes. Viewing Einstein's Cross is far more difficult, however – not only are the components very faint and closely separated, but they surround the dim core of the galaxy. Barbara Wilson has had some success viewing the brightest (A and B) components with a 20 in. scope, and has managed to barely 'resolve' the quasar with a 36 in. scope. Though a few observers have reported some success with scopes smaller than 20 in., it is advised to use the largest instrument available and at very high (over 500×) power before making the attempt.

Elongated and Edge-on Systems

Edge-on galaxies are most popular targets for the galaxy hunter [162]. The striking feature is their degree of elongation: often appearing as a long, thin luminous streak in the eyepiece – not comparable to any other object (see for instance NGC 5907 in Draco; Fig. 8.11). At high inclination many details become visible: dark lanes, sharp/diffuse

Table 8.14. The brightest variable galaxies, designated as "variable stars" (Dist = distance in Mpc, G = galaxy)

Object	Type	R.A.	Decl	Variation	z	Dist	Remarks
IO And	QSO	00 48 19.0	+39 41 09	15.3–17.6	0.134	533	
UX Psc	AGN	01 11 45.4	+22 04 10	13.4–16.3	0.0456	188	
BW Tau	AGN	04 33 11.1	+05 21 14	13.7–16.4	0.033	137	3C 120
CSV 6150 (Tau)	G	05 10 50.5	+16 28 33	11.7–16.5			CGCG 469–3
S10838 (Aur)	AGN	05 54 53.6	+46 26 22	14.4–15.5	0.0205	86	UGC 3374
GQ Com	QSO	12 04 42.1	+27 54 12	14.7–16.1	0.165	648	
X Com	AGN	13 00 22.5	+28 24 03	12.5–17.9	0.092	372	PGC 44750
AU CVn	QSO	13 10 28.6	+32 20 44	14.2–20.0	0.996	3158	
AP Lib	BL	15 17 41.9	–24 22 22	14.0–16.7	0.042	174	
V395 Her	AGN	17 22 34.1	+24 45 00	16.1–17.7	0.0638	261	8 Zw 476
V396 Her	QSO	17 22 41.2	+24 36 18	15.7–16.7	0.175	684	
V1102 Cyg	AGN	19 10 37.2	+52 13 13	15.5–17.	0.027	113	PGC 62859
V362 Vul	G	20 02 48.6	+22 28 27	16.0–17.7	0.029	121	

Fig. 8.9. The variable starburst galaxy V362 Vul

Table 8.15. Observations of variable galaxies

Object	Description
IO And	Directly visible, stellar (20″, 500×)
UX Psc	Directly visible, slightly nonstellar (20″, 312×)
BW Tau	Directly visible, almost stellar (14″, 250×)
	Easy, bright core with faint elliptical halo (20″, 312×)
CSV 6150 (Tau)	Directly visible, compact center with faint round halo (20″, 500×)
S10838 (Aur)	Easily visible, compact center, large faint oval halo (20″, 312×)
GQ Com	Averted vision, stellar (14″, 250×)
X Com	Directly visible, stellar (20″, 500×)
AU CVn	Averted vision, stellar (20″, 500×)
AP Lib	Directly visible, stellar (14″, 250×)
V395 Her	Directly visible, stellar (14″, 250×)
	Slightly diffuse (20″, 500×)
V396 Her	Averted vision, stellar (14″, 250×)
V1102 Cyg	Directly visible, stellar (20″, 500×)
V362 Vul	Averted vision, very small in crowded field (20″, 500×)

edges and/or the bulge. Generally systems with a high axis ratio (a/b) are called "flat galaxies." The most extreme case are the "superthin galaxies." The tables below show some ordinary edge-ons (Table 8.18) and a few superthin systems (Table 8.19), for a taste of the extreme. The prototype of a superthin galaxy is IC 2233, just south of

Table 8.16. Brighter lensed quasars (d = separation in arcsec, Dist = distance in Mpc)

Object	Con	R.A.	Decl	V	d	z	Dist	Name
APM08279+5255	Lyn	08 31 41.6	+52 45 18	16.5	0.4	3.870	9855	
Q 0957+561	UMa	10 01 20.9	+55 53 52	16.7/17.0	6.1	1.414	4225	Double Quasar
HE 1104–1805	Crt	11 06 33.6	−18 21 25	16.2	3.0	2.319	6375	
PG 1115+080	Leo	11 18 17.1	+07 46 01	15.8	2.1/2.7	1.724	4979	Triple Quasar
UM 425	Leo	11 23 20.7	+01 37 48	16.1	6.5	1.465	4351	
H 1413+117	Boo	14 15 46.3	+11 29 44	17.0	1.4	2.546	6896	Clover Leaf
Q 2237+0305	Peg	20 40 30.3	+03 21 30	16.8	1.8	1.695	4910	Einstein Cross

Table 8.17. Quasar pairs (*d* = separation in arcsec, Dist = distance in Mpc)

Object	Con	R.A.	Decl	*V*	*d*	*z*	Dist
Q 0023+171	Psc	00 25 37.0	+17 28 02	21.9/23.1	4.8	0.945	3023
CT 344	Scl	01 05 34.7	−27 36 59	17.8	0.3	0.848	2761
PHL 1222	Psc	01 53 53.9	+05 02 59	17.6/21.5	3.3	1.904	5408
CTQ 839	For	02 52 57.9	−32 49 09	18.3/20.8	2.1	2.240	6193
OM-076	Crt	11 47 52.6	−07 24 43	18.7/21.5	4.2	1.342	4046
HS 1216+5032	CVn	12 18 40.6	+50 15 37	17.2/18.6	8.9	1.450	4314
Q 1343+266	Boo	13 45 44.2	+26 25 07	20.0/20.4	9.5	2.030	5704
RIXOS F212–032	Her	16 29 02.4	+37 24 32	18.6/18.8	4.3	0.923	2964
FIRST J1643+3156	Her	16 43 11.4	+31 56 20	18.4/18.4	7.8	0.586	2019
LBQS 2153–2056	Cap	21 55 53.5	−20 41 46	17.9/21.3	3.0	1.845	5268
MGC 2214+3550	Lac	22 14 57.3	+35 51 26	18.8/19.3	3.0	0.877	2840

Fig. 8.10. The famous "Double Quasar" in Ursa Major

another strange object: NGC 2537 or "Bear Paw Galaxy." As described in Section II, the best technique to glimpse a superthin is slow field sweeping; remembering that most of these systems have a low surface brightness (see observations in Table 8.20). The opposite of these systems are face-on galaxies, but these would not be presented as a separate listing. However there are a number of prominent cases in the Messier catalogue, e.g., M 101, M 51, M 74, and M 83.

Fig. 8.11. NGC 5907 in Draco, a superb edge-on galaxy

Additional Notes

NGC 55. One of the brightest members of the Sculptor Group, this is a large Magellanic-type barred spiral (Fig. 1.12). Nearly as long as the Full Moon, its surface is strewn with bright knots, HII and other star forming regions. Had it, and many of the other group members, been located farther north, its likely they would have been included in Messier's catalogue.

NGC 100. A good example of a superthin galaxy, it is elongated nearly 10:1, though visually this is closer to 6:1. A dusky galaxy with no prominent dust lane or nuclear bulge, this system has proven to be quite difficult to classify accurately (Fig. 1.26).

NGC 891. A beautiful galaxy in Andromeda (Fig. 6.6), it along with NGC 4565 (Fig. 1.14) and NGC 4631 (Fig. 3.8) ranks as one of the finest edge-ons in the northern sky. The broad dust band is very complex, with numerous dark finger-like projections penetrating into the galaxy's disk. Called "chimneys," they are thought to be the result of supernova explosions expelling dust/gas deep into the starry disk. Visually in large scopes, the dust band appears quite scalloped and irregular, while hints of these structures are detectable at high magnification.

IC 2233. The "Needle" (Fig. 8.12), located about 15′ SSE of the peculiar galaxy NGC 2537, aka "Bear Paw Galaxy" (see Table 8.21). Though it shows signs of star forming activity, its disk is nearly dust-free and like most members of its class it has only a very weak nuclear bulge.

Table 8.18. Selected bright, large edge-on galaxies

Object	Con	R.A.	Decl	V	V'	$a \times b$	PA	Type	Remarks
NGC 55	Scl	00 15 08.0	−39 13 10	7.8	13.3	31.2×5.9	108	SB(s)m: sp	Sculptor group
NGC 891	And	02 22 33.0	+42 20 50	10.1	13.1	11.7×[illegible].6	22	SA(s)b? sp	
NGC 3109	Hya	10 03 06.6	−26 09 30	9.8	14.3	19.1×3.7	93	SB(s)m sp	DDO 236, beyond Local Group
NGC 4244	CVn	12 17 29.9	+37 48 28	10.0	13.1	16.6×1.9	48	SA(s)cd: sp	RFGC 2245
NGC 4517	Vir	12 32 45.6	+00 06 56	10.5	13.3	10.5×1.5	83	SA(s)cd: sp	RFGC 2315
NGC 4565	Com	12 36 20.5	+25 59 16	9.5	13.2	15.8×2.1	136	SA(s)b? sp	RFGC 2335
NGC 4631	CVn	12 42 07.6	+32 32 30	9.0	12.9	15.2×2.8	86	SB(s)d sp	Arp 281
NGC 4656	CVn	12 43 58.1	+32 10 11	10.1	13.9	10.0×1.8	37	SB(S)m pec	FGC 174A
NGC 4945	Cen	13 05 26.1	−49 27 46	8.6	13.2	19.8×4.0	43	SB(s)cd: sp	
ESO 270-17	Cen	13 34 48.2	−45 33 01	10.7	13.7	12.3×1.5	110	SB(s)m:	RFGC 2603, Fourcade-Figueroa
ESO 274-1	Lup	15 14 13.2	−46 48 45	10.8	13.6	10.5×1.5	36	SAd: sp	RFGC 2937
NGC 5907	Dra	15 15 53.8	+56 19 49	10.4	13.4	12.6×1.4	155	SA(s)c: sp	RFGC 2946
NGC 7640	And	23 22 06.6	+40 50 42	11.1	14.1	10.5×1.8	167	SB(s)c	

Table 8.19. Selected superthin galaxies (RFGC = Revised Flat Galaxy Catalog; $R = a/b$)

Object	RFGC	Con	R.A.	Decl	V	V'	$a \times b$	R	PA	Type
NGC 100	95	Psc	00 24 02.6	+16 29 11	12.7	14.1	6.16×0.64	9.6	55	Sc
UGC 711	255	Cet	01 08 37.0	+01 38 29	13.8	14.3	4.65×0.30	15.5	118	Scd
IC 2233	1340	Lyn	08 13 59.5	+45 44 23	12.3	13.1	5.17×0.58	8.9	173	Scd
UGC 7321	2246	Com	12 17 33.8	+22 32 29	13.4	13.7	5.54×0.36	15.4	81	Sd
NGC 5023	2495	CVn	13 12 11.8	+44 02 14	12.1	13.6	7.28×0.78	9.3	28	Scd
UGC 9242	2774	Boo	14 25 21.6	+39 32 18	13.5	13.7	5.66×0.34	16.6	71	Sd
UGC 9977	3021	Ser	15 42 00.0	+00 42 47	13.2	13.4	4.26×0.40	10.7	77	Sc

Table 8.20. Observations of edge-on and superthin galaxies (the galaxies NGC 55, NGC 3109, NGC 4565, NGC 4631, NGC 4945 were already described in Table 7.5)

Object	Description
NGC 100	Very faint, thin edge-on 6:1 WSW–ENE, moderately large, 2.0′×0.3′, weak concentration (SG 17.5″)
NGC 891	Fairly bright, large, edge-on, central bulge (SG 8″)
	Bright, extremely large, edge-on 5:1 SSW–NNE, 10′×2′. A striking dust lane bisects the galaxy and is most prominent through the bulging central region (SG 17.5″)
NGC 4244	Fairly bright, extremely large edge-on about 10:1 SW–NE. Extends to 15′–20′ diameter (fades at the ends of the extensions). Appears as a narrow ray with only a weakly concentrated core (SG 13″)
NGC 4517	Moderately bright, very large edge-on 8 in.: 1 WSW–ENE, almost 10′×1.2′. This galaxy is an impressive large narrow streak with fairly low surface brightness and fills 1/2 of the 21′ field, no sharp nucleus but central bulge. Appears brighter along the western extension (SG 17.5″)
NGC 4656	Striking! Fairly bright, very elongated SW–NE. Appears wider and brighter at the SW end. The NE end hooks sharply E to merge with NGC 4657, which may be a part of NGC 4656 and not a separate galaxy. A star or knot is attached at the S end. Appears like a celestial hockey stick! (SG 17.5″)
NGC 4945	A very large and impressive object – with a bright nuclear hub and numerous bright knots and dark dusty motes. A superb object that is somewhat reminiscent of NGC 253 (RJ 20″)
ESO 270-17	Only the SE third of this most peculiar object was visible. A very faint, diffuse streak measuring ~5′×2′, with slight condensations visible (RJ 18″)
NGC 5023	Fairly faint, pretty large, extremely long, weak center (14″)
NGC 5907	Very large, very elongated, narrow streak, bright core, faint star is W of the core (SG 13″)
	Fairly bright, extremely large edge-on 9:1 NNW–SSE, extends to roughly 13′×1.5′ Contains a bright core with an almost stellar nucleus. A 14 mag star lies 1.1′ W of center (SG 17.5″)
NGC 7640	Faint, large, very elongated streak N–S. There are faint stars at both the N and S end (SG 8″)
	Moderately bright, very large, very elongated 4:1 N–S, 7.0′×1.5′, large slightly brighter middle bulges. A 13.5 mag star is at the SE edge of the core 33 in. from the center. Bracketed by two 11 mag stars at the N end 3.0′ NNW of center and just W of the S end 2.6′ SSW of center. An extremely faint 15 mag star is embedded near the N end (SG 17.5″)
IC 2233	Very faint, extremely elongated, low surface brightness, very difficult (14″)
	Very faint, moderately large, extremely thin edge-on NNW–SSE with a low even surface brightness. A 14 mag star is embedded at the N tip and a 11/14 mag double star at 13 in. separation is off the E side 1.0′ from center (SG 17.5″)
UGC 711	Faint, extremely flat, weak center (20″)
UGC 7321	Fairly faint, extremely flat streak, no bulge (20″)
UGC 9242	Very faint, flattest system ever seen, difficult, very weak center (20″)
UGC 9977	Faint, very flat, low surface brightness (20″)

NGC 4244. A large, nearby galaxy that has a small core and a weak, disjointed equatorial dust band. While many edge-on galaxies have smooth disks, this system has a number of deformations or corrugations that have warped the disk [208]. This is thought to the result of an unusual type of density wave. NGC 5023 has a similar corrugated morphology.

UGC 7321. Called one of the "thinnest galaxies known" by some researchers, this galaxy is wispy streak that is elongated an astonishing 15:1! This low luminosity spiral has only a weak nuclear region and no visible bulge, even in the largest telescopes.

NGC 4945. A large, highly inclined barred spiral located deep in the constellation of Centaurus. The third brightest member of the Centaurus Group; other bright members include M 83 (Fig. 3.3) and NGC 5128 (Fig. 7.8), its deep southerly location has kept it from becoming a better known showpiece. This is the closest known Seyfert Type II galaxy and its bright core is thought to be powered by a massive black hole.

ESO 270-17. Sometimes known as the *Fourcade-Figueroa Object,* this is a galaxy 'shred' or shard. According to researchers, around 500 million years ago a spiral galaxy collided with the giant elliptical NGC 5128 (Centaurus A; Fig. 7.8). The orientation of both the dust band of NGC 5128 and the "shred", plus its peculiar lack of rotation suggests its collision based origin.

Fig. 8.12. A typical superthin galaxy: IC 2233 in Lynx

NGC 5907. The "Splinter Galaxy" (Fig. 8.11), it is one of the largest edge-on systems in the sky (inclination $i = 86.5°$). It has a small nuclear bulge and a slightly off-centered dust band that is most visible as it crosses the core region.

(For NGC 4565, NGC 4631 and NGC 4656 see "Additional Notes" after Tab. 7.5)

Peculiar and Amorphous Galaxies

Many "normal" galaxies can be found in the tables in each of the previous sections. Many would fine subjects for the visual study of galactic morphology, and especially the Hubble classification sequence [163,164]. Special and "unusual" types of galaxies have always fascinated both professionals and amateurs, and for many observers Halton Arp has become almost a "cult figure." The number of peculiar galaxies is very large, thus we can only present a small (and pretty subjective) sample here (Table 8.21). The focus is on single objects; double and multiple objects are featured below. Obviously these classes strongly overlap, often due to the phenomena of (tidal) interaction caught in various stages. Thus a "single object" may describe cases, where there is no companion (the interaction is long ago), or a merger is quite advanced, and the galaxies are already confined in a common envelope, like NGC 2623 or NGC 7252. The HST has imaged a lot of interesting peculiar and amorphous systems (Figs. 8.13 and 8.19).

Ring Galaxies

Ring galaxies are perhaps the oddest of all peculiar galaxies. They are the end result of a direct, central collision of two galaxies or the remnant of an ancient merger event [169]. Ring galaxies come in two distinct types – *classic* and *polar* rings (Table 8.22). Classic or equatorial rings like ESO 350-40, such as the well-known Cartwheel Galaxy are thought to be the result of an impact of a small, compact galaxy passing through the disk of a larger spiral more or less through the axis of rotation. The tidal disruption travels through the impacted system much like a series of ripples after a pebble is tossed into pond. The density waves propagate through the disrupted galaxy produced a ring-like structure of enhanced star formation.

Unlike the "classic" ring galaxy, polar rings are oriented perpendicular to the long axis of the galaxy. This type of ring is usually associated with a SO or "spindle shaped" galaxy, and thought to be remnants of an ancient galaxy merger event. Perhaps the finest examples in the sky are NGC 2685 in Ursa Major and the spectacular NGC 4650A in Centaurus (Fig. 3.5). Both types of ring galaxies are very rare – much less than 1% of all galaxies display this unusual and intriguing structure. Many of these galaxies are quite faint, making them "trophy objects" for even the advanced observer or astroimager (Table 8.23).

Additional Notes

ESO 350-40. The "Cartwheel" or sometimes known as "Zwicky's Ellipse." One of the best known of all ring galaxies, this disturbed system lies over 500 million ly distant. Spectacularly imaged by the HST, the bright ring measures over 150,000 ly in diameter and is undergoing an intense phase of star formation.

Fig. 8.13. NGC 6745, a peculiar starburst galaxy in Lyra

NGC 520. A highly unusual galaxy distorted by a 3-galaxy interaction. The twisted/distorted disk represents the partial merger of two galaxies, while the long tidal plumes may be the result of a close passage of UGC 957, a smaller satellite system.

NGC 985. A Seyfert Type I galaxy with a lopsided ring structure. One of the easiest rings to resolve, it can be glimpsed with telescopes as small as 16″.

NGC 2537. The "Bear Paw Galaxy," also called Mrk 86 and Arp 6. This is a good example of a blue compact dwarf (BCD). Such galaxies are undergoing intense and/or prolonged periods of intense star formation. Other well-known BCDs include ESO 495-24, NGC 1705, NGC 3125, and NGC 6789.

NGC 2685. The "Helix Galaxy," perhaps the finest polar ring galaxy visible in the northern sky. This name derives helical gas and dusty filaments that form a broad ring perpendicular to the long axis of the SO galaxy [169]. This polar ring is thought to be the remnants of an ancient merger event some 2 to 5 billion years ago.

NGC 4650A. Easily the most photogenic of all polar ring galaxies, the HST has beautifully imaged this object (Fig. 3.5). The ring is fairly young – an estimated 1–3 billion years old, and is undergoing a period of massive star formation. This object is a member of the large Centaurus Galaxy Cluster.

NGC 4861. This is an example of an exotic type, known as "cometary" dwarf galaxy. The largest HII region is designated as IC 3961 and forms the "tail," while the condensed "head" of this "intergalactic comet" is NGC 4861.

Table 8.21. Collection of peculiar or amorphous galaxies (due to irregular shape and light distribution no V' and PA are given)

Object	Con	R.A.	Decl	V	$a \times b$	Type	Remarks
NGC 454	Phe	01 14 21.6	−55 24 02	12.2	1.7×1.5	S pec	
NGC 520	Psc	01 24 34.7	+03 47 39	11.3	3.4×1.7	Sa pec	Arp 157
NGC 2537	Lyn	08 13 14.4	+45 59 29	11.7	1.7×1.5	SB(s)m pec	Bear Paw Galaxy, Arp 6
NGC 2623	Cnc	08 38 24.1	+25 45 17	13.2	2.4×0.7	Sb pec	Arp 243
UGC 5938	Dra	10 51 50.1	+77 34 19	15.6	1.0×0.2		VII Zw 349, "comet," 1st of 2
UGC 5942	Dra	10 51 59.6	+77 32 50	16.0	0.8×0.2		VII Zw 349, "comet," 2nd of 2
Arp 148	UMa	11 03 53.2	+40 50 57	15.0	0.8×0.5		Mayall's Object
NGC 3509	Leo	11 04 24.4	+04 49 42	13.0	2.1×1.0	SBbc pec	Arp 335, one-armed spiral
NGC 4774	CVn	12 53 06.6	+36 49 08	14.3	0.6×04	Ring	Kidney-Bean Galaxy, I Zw 45
HZ 46	CVn	12 56 55.6	+32 26 51	15.0	0.7×0.4		Mrk 54, 2 wings
NGC 4861	CVn	12 59 01.8	+34 51 43	13.5	4.2×1.6	SBm	IC 3961, Arp 266, "comet"
IC 883	CVn	13 20 35.5	+34 08 19	13.9	1.4×0.7	Im pec	Arp 193, 2 jets
UGC 9562	Boo	14 51 14.4	+35 32 32	13.9	1.1×1.1		VV 324, cross-shaped
II Zw 73	Boo	15 16 00.1	+43 09 46	15.5	1.0×0.2		Saturn-shaped
IC 1182	Her	16 05 36.7	+17 48 10	14.3	1.0×0.5	SA0+ pec	Arp 172, Hercules Cluster
UGC 10214	Dra	16 06 03.9	+55 25 32	14.4	3.6×0.8	SB(s)c pec	Tadpole, Arp 188
UGC 10491	Her	16 38 13.9	+41 56 06	14.3	1.1×0.5		Arp 125, "comet"
I Zw 207	Dra	18 31 10.4	+55 16 32	15.3	1.7×0.5		boomrang-shaped
NGC 6745	Lyr	19 01 41.6	+40 44 45	13.9	1.3×0.5	Sm pec	
UGC 11916	Peg	22 08 21.9	+18 27 14	15.0	1.0×0.6	S?	II Zw 166, "sandwich"
NGC 7252	Aqr	22 20 44.8	−24 40 42	11.1	2.7×1.7	SB0	Arp 226
NGC 7592	Aqr	23 18 22.0	−04 24 59	13.5	1.0×0.9		VV 731, double system
NGC 7732	Psc	23 42 34.0	+03 43 30	13.6	2.0×0.6	Scd pec	Zwicky's "pierced" galaxy

Table 8.22. A collection of the better known classic, polar, and irregular ring galaxies

Object	Con	R.A.	Decl	V	$a \times b$	Type	Remarks
ESO 350-40	Scl	00 37 41.1	−33 42 59	14.5	1.1×0.9	Classic	Cartwheel, Zwicky's Ellipse
NGC 660	Ari	01 43 02.0	+13 38 39	12.0	8.3×3.1	Polar	
UGC 1775	Cet	02 18 26.4	+05 39 12	13.8	1.5×1.4	Classic	Arp 10
NGC 985	Cet	02 34 37.4	−08 47 06	13.5	1.0×0.9	Classic	
NGC 1143/4	Cet	02 55 09.8	−00 10 41	14.1	0.9×0.7	Irregular	Multigalaxy, irregular ring, Arp 118
II Zw 28	Ori	05 01 14.9	+03 34 24	15.0	0.3×0.3	Classic	VV 790
NGC 2685	UMa	08 55 34.9	+58 44 05	11.2	4.6×2.5	Polar	Helix Galaxy, Arp 336
NGC 3661	Cra	11 23 38.5	−13 49 53	14.8	1.5×0.5	Polar	IC 689
UGC 7576	Com	12 27 41.8	+28 41 52	15.0	1.5×0.5	Polar	
UGC 7683	Uma	12 32 04.8	+66 24 11	15.5	0.4×0.3	Classic	VII Zw 466
NGC 4650A	Cen	12 44 49.0	−40 42 52	13.3	1.6×0.8	Polar	
NGC 4774	CVn	12 53 06.6	+36 49 08	14.3	0.6×0.4	Classic	
Abell 76	Hya	21 30 04.7	−02 48 22	14.5	0.4×0.4	Classic	Misclassified PN

Table 8.23. Visual descriptions of peculiar, ring and amorphous galaxies

Object	Description
NGC 454	Slightly faint 1′×0.5′ haze, elongated E–W, which rises to a small elongated core. The halo is extended on the W side and touches ESO 151-36A, a faint 0.7′ diameter irregularly-round haze which is broadly brighter to the centre. A 11 mag star is 1.5′ N of the pair. Best viewed at 450× (Michael Kerr 25″)
NGC 520	Faint, diffuse, elongated N–S (SG 8″)
	Fairly bright, moderately large, elongated 5:2 NW–SE, 3.0′×1.2′. Very unusual appearance; the NW portion is noticeably brighter with a bright knot at the NW tip and a mottled texture. Fades toward the SE where it merges into a fainter section which is tilted ~E–W with an irregular surface brightness and ill-defined edges (SG 17.5″)
	An irregular dust lane cuts across the roughly polygonal shaped disk, with faint wispy tidal plumes visible off the ends of the bright disk (RJ 24″)
NGC 660	Fairly large and diffuse, with only a modest core concentration. The disk is oriented NE–SW, and the surface is irregularly mottled with dusty streaks (RJ 20″)
NGC 985	Fairly faint, very small, round, 15 in. diameter, sharp stellar nucleus with a small very faint halo! A triangle of 10/11 mag stars with sides 1.7′, 2.5′, and 3.0′ is about 5′ WNW and the galaxy forms the bottom of a "cross" asterism with these stars (SG 17.5″)
NGC 2537	Fairly faint, small, round, no structure (SG 13″)
	Moderately bright, fairly small, round. Unusual appearance as there is a dark lane or vacuity in the center. A small slightly brighter knot is visible along the NW edge (SG 17.5″)
NGC 2623	Faint, small, slightly elongated, weak concentration (SG 13″)
NGC 2685	Moderately bright, fairly small edge-on 4:1 SW–NE. Contains an elongated bright core. A 11 mag star is 2.4′ N of center (SG 13″)
	Traces of the dusty bands cutting across the center of the spindle are visible under high magnification (RJ 24″)
NGC 3509	Fairly faint, moderately large, elongated 5:2 SSW–NNE, 1.6′×0.7′. Low surface brightness with a very weak concentration (no visible core). Difficult to determine outer extent of halo but appears to have an asymmetric shape (slightly curved?) (SG 17.5″)
NGC 4650A	I could barely identify NGC 4650A which has the appearance of a very faint, out-of-focus little star. Had I not known that it should be exactly there, I would never have seen it (Magda Streicher 12″)
	Faint 0.5′×0.2′ haze, elongated ENE–WSW, which rises to a stellar nucleus. Averted vision shows a very faint 1′-long extension to the NNW and an even fainter extension to the SSE. A 16 mag star is 1′ WSW and two 17 mag stars are 0.2′ E and 0.6′ SE. Best viewed at 450× (Michael Kerr 25″)
NGC 4774	Very faint, small, round, even surface brightness (SG 17.5″)
NGC 4861	Faint, very elongated SSW–NNE, even low surface brightness. Located between two 12 mag stars at low power. This "star" is slightly nebulous at 166× and a definite nonstellar knot is visible at 312×. This is one of the few extragalactic HII regions which responds to OIII filtration (SG 13″)
NGC 6745	Fairly faint, edge-on SSW–NNE. At 220× appears to bend on the NNE end to the W. Extension seen at the NNE end may be a contact pair (SG 13″)

(*Continued*)

Table 8.23. Visual descriptions of peculiar, ring and amorphous galaxies—Cont'd

Object	Description
NGC 7252	Fairly faint, very small, round, compact, weak concentration (SG 13″)
NGC 7592	Faint, small, round. Just resolved is a very faint and extremely small companion (only nucleus observed) attached at the W edge (SG 17.5″)
NGC 7732	Very faint, fairly small, elongated 2:1 E–W, low even surface brightness. Located just 1.0′ S of a 11 mag star. Forms a close pair with NGC 7731 1.4′ NW (SG 17.5″)
IC 883	Fairly bright, oval, diffuse, brighter to the middle, no jets visible (20″)
IC 1182	Very faint, very small, slightly elongated. Situated between two 14.5 mag stars 1.4′ W of center and a 15 mag star following (SG 17.5″)
UGC 7576	Faint, very small, round (20″)
UGC 9562	Very faint, irregular shape (20″)
UGC 10214	Pretty faint, elongated, brighter middle, no sign of steamer (20″)
UGC 10491	Fairly faint, pretty elongated, asymmetric core (20″)
UGC 11916	Pretty faint, small, slightly oval halo (20″)
ESO 350-40	Very difficult object. Appears as an extremely faint 1.3′ × 1′ disk, elongated SE–NW close to a 15 mag star 2′ NW and a 14 mag star 3′ WNW. It is much fainter than expected and the visibility is affected by the transparency. There may be glimpses of a brighter rim to the disk once but this is probably imagination. No other structure is visible. The companions are possibly visible with averted vision as a single glow but they are at the limit and cannot be held (Micheal Kerr 8″)
	Visible as a faint 1.3′ × 1′ oval haze elongated NW–SE with a brighter broken rim along the SW side and a 16 mag star just off the W edge of the rim. Within the rim averted vision shows a slight but definite brightening, which has an uneven surface brightness and appears darker on the SW side, and a faint core. ESO 350-40B appears as a 15 mag stellar nucleus with an extremely faint 15 in. diameter halo, which is only really seen with averted vision. Best viewed at 350× or 450× (Michael Kerr 25″)
Arp 148	Faint, elongated shape, diffuse (20″)
I Zw 207	Very faint, like a small arc, difficult (20″)
II Zw 28	Very faint, round (14″)
HZ 46	Pretty faint, compact, slightly diffuse (14″)
Abell 76	Pretty faint, large diffuse disk of homogenous brightness (20″)

NGC 6745. A triple galaxy system, in which the smaller components has collided with a much larger spiral galaxy (Fig. 8.13). As collisions go, this one is fairly recent – only a few hundred million years old.

NGC 7252. The "Atoms for Peace" galaxy, this peculiar elliptical is the result of a merger of two large spiral galaxies over a period of a billion years ago. Long tidal tails are still evident, and HST images have revealed that this system is the host for more than 500 SSCs (see Waller & Hodge). Perhaps in 4 or 5 billion years in the future, M 31 and our own Milky Way will produce a similar looking merged system.

IC 883. Also known as Arp 193, this is an infrared luminous example of a galactic merger. It has a bright starburst nucleus and two tidal tails projected at nearly right angles from the disk of the system.

IC 1182. A brilliant starburst galaxy (Arp 172) located near the center of the Hercules Galaxy Cluster. This is considered an on-going merger between a giant elliptical and

Fig. 8.14. A cosmic sabre: UGC 10214 in Draco, also know as "Tadpole"

a smaller spiral galaxy. Several jets and gaseous arcs are visible in deep images, as first noted by Arp in 1972.

UGC 10214. The "Tadpole" or Arp 188 (Fig. 8.14). This galaxy has an incredibly long, thin tidal tail. Numerous HII regions and SSCs are strewn along this tail, as evident in the HST images. This tail is reminiscent of similar highly elongated tidal tails that are associated with the "Antennae" (NGC 4038/9, Fig. 1.35) and the Mice (NGC 4676, Fig. 8.15).

Monsters in the Dark: Giant Ellipticals and cD Galaxies

Giant ellipticals and their close cousins the cD galaxies are the most massive and luminous class of galaxies [209]. The largest members can tip the scales at well over 10 trillion solar masses. Generally these immense objects reside near the centers of galaxy groups and clusters. Their generally round or ellipsoidal appearance and smooth light distribution lends to the assumption that these are inactive systems comprised of primarily

Fig. 8.15. "The Mice" = NGC 4676 in Coma Berenices

ancient stars – they are not! These are deceptively complex systems. Some are spheres, others oblate or even triaxial in shape and can range from nearly nonrotating to fast rotating systems. The isophotes of the inner structures can range from spheroidal to "disky" to even boxy or rectangular.

Relatively rare and even more massive are the cD galaxies [210]. These are giant ellipticals with extensive diffuse halos that can encompass much of the central region of a large cluster (cD = "core dominant"). These galaxies are found almost exclusively in the center of large galaxy clusters and are often surrounded by a number of smaller systems.

Both giant ellipticals and cD galaxies will tidally disrupt and merge with smaller objects in a process known as galactic cannibalism. Well known objects such as Centaurus A, M 87, and even our own Milky Way have taken their toll on their smaller neighbors. Many cD galaxies are multinucleated as they can have several nuclear concentrations in their cores. The distant giant NGC 6166 (Fig. 1.22) has at least three cores and is also surrounded by a host of other small galaxies and represents an extreme example of cannibalism.

In some large galaxy clusters one can observe paired giant ellipticals. Nicknamed "dumbbell galaxies" from their apparent shape, these galaxies rotate around a common center of gravity and are so close that they can be immersed in a common outer halo or envelop. Most are extremely massive systems, often tipping the scales well in excess of 10^{13} solar masses [216]. These pairings are not uncommon, and can be found in many dense galaxy clusters.

The descriptions and technical data for a number of giant ellipticals can be found in the Messier Catalogue (M 49, M 59, M 60, M 84, M 86, M 87, M 89, and M 105) in Tables 7.1 and 7.2. Observations and data on other interesting giant ellipticals, cD, and dumbbell galaxies can be found in Tables 8.24 and 8.25.

Additional Notes

NGC 545/7. A pair of corotating, giant ellipticals of the "dumbbell galaxy" class. Both systems are surrounded by a large, common halo. The smaller, more condensed galaxy (NGC 547) is also a radio source (3C 40).

NGC 1275. The origin of its unusual shape and dusty motes of this object has been a subject of speculation for many decades. But recent high-resolution images from the HST have finally settled this controversy. The images reveal a large dusty spiral colliding with the giant cD galaxy. In the next 100 million years or so, this galaxy will be torn to shreds and incorporated into the much more massive elliptical.

NGC 1399. Located near the heart of the Fornax Cluster, this cD galaxy has a huge globular cluster population. There seems to be a fairly good correlation between the number of GCs and the mass of a galaxy. While over 150 have been found for the Milky Way – at least twice many have been found associated with M 31 [207]. For the massive NGC 1399, nearly 6,000 GCs have been detected. Even larger populations have been found with M 87 (+10,000) and other giant elliptical and cD galaxies [209]. Tidal interactions and mergers with other galaxies have had a profound impact on the globular cluster populations associated with these galaxies.

NGC 3923. Deep exposures of this galaxy have revealed a series of faint interleaved arc-like shells distributed symmetrically around the main disk of the galaxy [209]. These shells are thought to be an outward moving density wave as the result of a merger with a smaller galaxy [211]. Like other large ellipticals, this system has a large population of GCs orbiting the main disk.

NGC 4782/3. Often considered the archetype of a dumbbell galaxy pair, these giant ellipticals are located in the center of a small cluster. Both objects are much larger and far more massive than our own Milky Way, tipping the scales at upwards of 10^{13} solar masses [216].

NGC 6166. Lying near the center of A 2199, this supergiant cD galaxy (Fig. 1.22) is one of the most massive and brightest galaxies known [212]. It is the archetype of the multiple-nucleus galaxy, as it has several – the remnants of past acts of cannibalism. It has a huge number of GCs in orbit, and also hosts a gigantic black hole in its center.

NGC 6240. A superluminous galaxy in the middle of a merger event (Fig. 1.31) – it is a spectacular example of an "Ultra Luminous Infrared Galaxy" (ULIRG). With a bolometric luminosity of over 2.4 trillion suns, it is over 50× brighter than our own Milky Way [213]. In several 100 million years it will settle down to become a giant elliptical or cD galaxy.

UGC 9799. A distant giant elliptical galaxy in A 2052, it has one of the largest populations of GCs ever recorded. With an estimated 46,000 GCs – this is at least 4 times that of M 87 or about 300 times of our own galaxy.

Table 8.24. Selected giant elliptical, cD, and dumbbell galaxies

Object	Con	R.A.	Decl	V	V'	$a \times b$	PA	Type	Remarks
NGC 545	Cet	01 25 59.1	−01 20 26	12.2	13.5	2.4×1.5	63	SAO-E1	Dumbbell galaxies, Arp 308
NGC 547		01 26 00.7	−01 20 42	12.2	12.7	1.0×0.9	85		
NGC 750	Tri	01 57 32.7	+33 12 32	12.0	12.9	1.6×1.3	162	E pec	Dumbbell galaxies, Arp 166
NGC 751	Tri	01 57 33.0	+33 12 10	12.2	12.7	1.2×1.2		E pec	
NGC 1275	Per	03 19 48.2	+41 30 42	11.9	13.2	2.2×1.7	103	cD	Perseus A Cluster
NGC 1399	For	03 38 29.0	−35 26 58	9.6	13.7	6.9×6.5	105	E pec	Fornax Cluster
NGC 3923	Hya	11 51 01.5	−28 48 19	9.8	13.2	5.8×3.8	50	E4-5	Numerous shells
NGC 4261	Vir	12 19 23.2	+05 49 29	10.4	13.3	4.3×3.5	150	E2-3	Virgo Cluster
NGC 4365	Vir	12 24 28.2	+07 19 03	9.6	13.5	6.9×5.0	40	E3	Virgo Cluster
NGC 4782	Crv	12 54 35.8	−12 34 11	11.7	12.9	1.8×1.7	155	E0 pec	Dumbbell galaxies
NGC 4783		12 54 36.4	−12 33 29	11.6	12.2	1.3×1.3		E0 pec	
NGC 4874	Com	12 59 35.7	+27 57 33	11.7	13.1	2.3×2.3		cD	Coma Cluster
NGC 4881	Com	12 59 57.8	+28 14 47	13.6	13.6	1.0×1.0		E+	Coma Cluster
NGC 4889	Com	13 00 08.1	+27 58 36	11.5	13.3	2.8×2.2	82	cD	Coma Cluster
NGC 5018	Vir	13 13 00.9	−19 31 04	12.8	13.5	1.7×1.3	102	E3	
NGC 5629	Boo	14 28 16.3	+25 50 55	12.1	13.2	1.8×1.8		S0/cD	Abell 2666 cluster
NGC 6166	Her	16 28 38.5	+39 33 05	11.8	12.8	2.2×1.5	32	CD; pec	Multinucleated
NGC 6240	Oph	16 52 58.8	+02 24 11	12.9	13.7	2.1×1.0	27	I0: pec	ULIRG, merging system
NGC 7619	Peg	23 20 14.5	+08 12 23	11.0	12.9	2.5×2.3	30	E	Pegasus I Cluster
NGC 7626	Peg	23 20 42.4	+08 13 02	11.1	13.0	2.6×2.3	19	E pec	Pegasus I Cluster
UGC 9799	Ser	15 16 44.6	+07 01 17	13.0	13.5	1.8×0.9	36	E	Abell 2052 Cluster

Table 8.25. Visual descriptions of giant elliptical and cD galaxies

Object	Description
NGC 545/547	Two bright elliptical galaxies that are nearly touching. NGC 545 is decidedly ellipsoidal and with condensed core, while NGC 547 is nearly round, smaller and has a higher surface brightness. Both objects located near the center of a rich Abell galaxy cluster A 194 (RJ 20″)
NGC 750/1	Double galaxy, N–S, two distinct nuclei in a common halo (SG 13″) Resolved into two distinct galaxies at 220×. NGC 750 moderately bright, small and round. Forms a contact double system with NGC 751, virtually attached to the south end. Fairly faint, very small, and round. Appears smaller and fainter than NGC 750 (SG 17.5″)
NGC 1275	The brightest member of the Perseus Cluster (A 426), it lies in a field crowded with galaxies. NGC 1275 appears as irregularly oval, with a bright, nearly stellar core. The halo appears to be oriented nearly E–W, and a faint star appears about 1′ to NE of the core (RJ 24″)
NGC 1399	Fairly bright, small, round, right core (SG 8″) Bright, large faint halo is broadly concentrated, brighter core. A star is superimposed 0.3′ N of the center (SG 13″)
NGC 3923	Bright, moderately large, elongated SW–NE, small bright nucleus (SG 8″) Bright, elongated, pretty large – adverted vision helps (SG 17.5″)
NGC 4365	Pretty bright, pretty large, little elongated 1.5′×1′, much brighter core – adverted vision makes the object appear larger (SG 11″) Very bright, large, elongated SW–NE, bright core – nearly stellar nucleus (SG 17.5″)
NGC 4782/4783	Two nearly equal sized objects, round, and in apparent contact. Both have nearly the same brightness and are oriented N–S (RJ 20″)
NGC 4874 NGC 4889	The giant "eyes" of the Coma Cluster. Both objects are much larger and brighter than the surrounding galaxies. NGC 4874 is nearly round, has a bright condensed core. NGC 4889 is markedly oval, and is elongated nearly E–W. Like its neighbor, it has a bright, condensed core. Both systems are surrounded by large numbers of smaller objects, esp. NGC 4874 (RJ 24″)
NGC 6166	Irregularly oval, fairly small object surrounded by a number of very faint galaxies. The core region is brightly mottled and unusual, as there appears to be several small knots visible (467×). The diffuse, outer halo is elongated from the NNW to SSE (RJ 20″)
NGC 6240	Fairly faint, small, elongated 2:1 SW–NE, even surface brightness. A 13 mag star is at the NE edge 0.6′ from the center (SG 17.5″)
NGC 7619 NGC 7626	Both objects are very similar in appearance, as they are quite bright, nearly round, diffuse and have a bright, well-condensed core. NGC 7619 is slightly oval, other wise it is a near twin of the other galaxy (RJ 20″)

Chapter 9

Groups and Clusters of Galaxies

Pairs and Trios

The Messier catalogue is a good place to start as it contains some pairs and trios of galaxies (see Table 7.1): M 81/M 82, M 95/M 96, M 84/M 86, M 65/M 66 (forming the "Leo Triplet" with the dusty edge-on NGC 3628), and the trio M 31/M 32/M 110. More than 300 examples have been found, where two or more galaxies brighter than 14 mag are within a field of 15′ – ideal targets for an 8″ telescope at medium magnification. For the northern hemisphere ($\delta > -20°$) 215 are listed in the *Catalog of Galaxy Groups* [165]. For the whole sky check the *Atlas of Galaxy Trios* by Miles Paul [166] as it presents 560 cases. These catalogues are designed for amateur use and ignore the question of chance alignment or interaction – their aim is simply presenting nice views in the eyepiece. See also the collection of Al Lamperti, which describes visual observations [247].

Respectively, 10 pairs and 10 trios were selected in the tables below (Tables 9.1 and 9.2; visual descriptions in Tables 9.3 and 9.5). There are a lot more cases; some that are quite well known (but pretty faint), like "Wild's Triplet" in Virgo (Arp 248 at 11 46.7 −03 50) [167] or "Zwicky's Triplet" in Hercules (Arp 103, 16 49.5 +45 29). In many cases tidal interaction is present, and often visible. The observation of interacting galaxies, e.g., those shown in Arp's atlas, is an interesting task [168]. One can see all kinds of tidal distortions: bridges, tails, and rings [169], which can look quite dramatic. Many of the more famous examples were also imaged by the HST (Fig. 8.15). In addition to the pairs and trios, we have included a small subset of superimposed galaxy pairs (Table 9.4). These are not true gravitational bound systems, but rather these systems are in the line of sight. Recently, these and other galaxies have studied by the HST and other large scopes to provide direct measurements of the effective absorption of the galaxy disks [214].

Let's finally mention one of the most extreme trios: II Zw 99 (PGC 66119) at 21 06 49.5 −00 51 28. It is the only known case of compact galaxies forming an equilateral triangle (Fig. 9.1). Unfortunately the object is not suitable for visual observing as the components are below 18 mag, plus located in a circle of 10″ across.

Table 9.1. Ten pairs of galaxies (d = separation of components in arcmin)

Object	Con	R.A.	Decl	V	V'	$a \times b$	PA	Type	d	Remarks
IC 1727	Tri	01 47 30.0	+27 19 57	11.4	14.1	5.7×2.4	150	SBm	8.4	VV 338
NGC 672	Tri	01 47 54.0	+27 25 58	10.7	13.4	6.0×2.4	65	SBc		VV 338
ESO 60–26	CAR	09 04 01.6	−72 03 18	13.6	12.8	0.9×0.6	135	Ring	0.8	
ESO 60–27	CAR	09 04 08.3	−72 03 01	13.6	13.0	0.9×0.7	120	Ring		
NGC 2992	Hya	09 45 41.9	−14 19 37	12.2	13.4	3.7×0.9	15	Sa	2.9	Arp 245
NGC 2993	Hya	09 45 48.3	−14 22 08	12.6	12.6	1.3×0.9	95	Sa		Arp 245
NGC 4038	Crv	12 01 52.8	−18 51 52	10.3	12.1	3.4×1.7	94	SBm	0.9	Arp 244, Antennae
NGC 4039	Crv	12 01 53.8	−18 53 08	10.4	12.1	3.3×1.7	55	SBm		Arp 244, Antennae
NGC 4085	UMa	12 05 22.4	+50 21 12	12.5	12.7	2.8×0.8	78	SBc	11.4	
NGC 4088	UMa	12 05 34.6	+50 32 26	10.3	12.8	5.6×2.1	43	SBbc		Arp 18
NGC 4298	Com	12 21 32.9	+14 36 24	11.4	13.2	3.2×1.9	140	Sc	2.4	
NGC 4302	Com	12 21 42.2	+14 35 54	11.9	13.6	5.3×1.0	178	Sc		
NGC 4676A	Com	12 46 10.1	+30 43 57	13.5	13.2	1.4×0.6	0	SB0-a	0.5	Arp 242, The Mice
NGC 4676B	Com	12 46 11.2	+30 43 21	13.8	14.3	2.2×0.8	2	S0-a		Arp 242, The Mice
NGC 5544	Boo	14 17 02.4	+36 34 16	13.3	13.3	1.1×1.0	62	SB0-a	0.6	Arp 199
NGC 5545	Boo	14 17 04.8	+36 34 29	15.0	13.5	1.0×0.3	58	Sbc		Arp 199
ESO 138–29	Ara	17 29 09.6	−62 26 44	11.7	12.9	2.5×1.4	45	Ring	3.0	
ESO 138–30	Ara	17 29 25.3	−62 28 50	14.0	13.3	1.2×0.5	145	SBab		
UGC 12914	Peg	00 01 38.3	+23 28 59	12.4	13.4	2.5×1.2	160	Scd:	1.1	VV 254
UGC 12915	Peg	00 01 42.2	+23 29 40	13.0	12.5	1.4×0.5	137	S?		VV 254

Table 9.2. Ten galaxy trios (components ordered by R.A.; AB, BC = separations in arcmin)

Object	Con	R.A.	Decl	V	V'	$a \times b$	PA	Type	Remarks
NGC 168	Cet	00 36 38.6	−22 35 37	14.0	12.8	1.2×0.3	26	S0-a	AB 8.0
NGC 172	Cet	00 37 13.6	−22 35 12	13.6	12.9	2.0×0.3	12	SBbc	BC 5.6
NGC 177	Cet	00 37 34.3	−22 32 57	13.3	13.3	2.2×0.5	11	Sab	
NGC 467	Psc	01 19 10.1	+03 18 05	12.1	13.1	1.7×1.7		S0	AB 10.9
NGC 470	Psc	01 19 44.8	+03 24 33	11.7	13.3	2.9×1.7	155	Sb	Arp 227, BC 5.8
NGC 474	Psc	01 20 06.7	+03 24 58	11.3	15.3	7.1×6.3	75	S0	Arp 227
NGC 1618	Eri	04 36 06.6	−03 08 55	12.7	13.3	2.4×0.8	26	SBb	Near Ny Eri, AB 8.0
NGC 1622	Eri	04 36 36.6	−03 11 18	13.1	14.0	3.7×0.7	33	SBab	Near Ny Eri, BC 10.0
NGC 1625	Eri	04 37 06.3	−03 18 14	12.3	12.2	2.1×0.5	131	SBb	Near Ny Eri
NGC 2295	CMa	06 47 23.2	−26 44 10	12.8	12.9	2.1×0.6	46	Sab	VV 178, AB 4.0
NGC 2292	CMa	06 47 39.4	−26 44 47	10.8	13.5	4.0×3.5	124	SB0	VV 178, BC 1.0
NGC 2293	CMa	06 47 42.8	−26 45 17	11.2	13.8	4.0×3.2	125	SB0-a	VV 178
M 105	Leo	10 47 49.5	+12 34 52	9.5	13.1	5.3×4.8	71	E1	AB 7.3
NGC 3384	Leo	10 48 16.7	+12 37 43	9.9	12.9	5.4×2.7	53	E/SB0	BC 6.5
NGC 3389	Leo	10 48 28.0	+12 31 59	11.8	13.1	2.9×1.3	112	Sc	
NGC 3991	UMa	11 57 30.7	+32 20 08	13.0	11.8	1.3×0.3	33	Im pec	Arp 313, AB 3.8
NGC 3994	UMa	11 57 36.8	+32 19 39	12.7	11.7	0.9×0.5	10	Sc pec	Arp 313, BC 1.9
NGC 3995	UMa	11 57 44.0	+32 17 35	12.1	12.9	2.6×0.9	33	SBm	Arp 313
NGC 4206	Vir	12 15 16.7	+13 01 22	12.0	14.0	6.4×1.1	0	Sbc	AB 11.3
NGC 4216	Vir	12 15 54.0	+13 08 52	10.3	13.1	8.1×1.8	19	SBb	BC 11.4
NGC 4222	Com	12 16 22.6	+13 18 25	13.2	13.5	3.1×0.5	56	Scd	
ESO 221–12	Cen	13 51 32.3	−48 04 57	13.4	12.9	1.5×0.5	164	SBm?	AB 10.0
ESO 221–13	Cen	13 51 35.1	−48 01 35	12.7	13.2	1.6×1.1	103	SB0-a	BC 3.4
ESO 221–14	Cen	13 52 07.1	−48 10 13	12.5	13.2	1.7×1.3	42	SBc	
NGC 6769	Pav	19 18 22.8	−60 30 03	11.6	12.7	2.2×1.5	123	SBb pec	VV 304, AB 2.0
NGC 6770	Pav	19 18 37.0	−60 29 46	12.0	13.2	2.2×1.6	20	SBb pec	VV 304, BC 3.3
NGC 6771	Pav	19 18 39.4	−60 32 47	12.6	12.6	2.3×0.5	118	SB0-a	
NGC 7232	Gru	22 15 37.6	−45 51 01	12.0	12.9	2.6×1.0	99	SBa	AB 2.0
NGC 7233	Gru	22 15 49.0	−45 50 47	12.2	12.9	1.7×1.3	133	SB0-a	BC 4.0
NGC 7232B	Gru	22 15 52.4	−45 46 50	13.0	13.8	1.6×1.5	0	SBm	

Table 9.3. Visual descriptions of galaxy pairs

Pair	Description
IC 1727 NGC 672	IC 1727 is very faint, moderately large, diffuse, ill-defined, elongated NNW–SSE, no central condensation (SG 13″). Very faint, moderately large, elongated 2:1 NW–SE. Very low surface brightness with no distinct edges or core (SG 17.5″) NGC 672 is fairly faint, low even surface brightness, fairly large, diffuse. Two 13.5 mag stars lie NW and at the E edge (SG 8″). Fairly bright, elongated 5:2 WSW–ENE, even surface brightness. Bracketed by a 13.5 mag star 2.2′ WNW and a 13 mag star 3.2′ E (SG 17.5″)
NGC 2207 IC 2163	Moderately bright, moderately large, bright core, double nuclei. A faint extension is visible to the E. A double nucleus is visible and an extension just seen to the E is probably IC 2163 (SG 13″)
ESO 60-26 ESO 60-27	ESO 60–26 is a faint 0.7′×0.3′ haze, elongated SE–NW, which rises to an elongated core and a stellar nucleus is visible with averted vision. A 15 mag star is off the NE end. ESO 60–27 is a faint 0.6′ round haze, which rises smoothly to a stellar nucleus. Best viewed at 450× (Michael Kerr 25″)
NGC 2992 NGC 2993	NGC 2992 moderately bright, small, slightly elongated SSW–NNE, bright core. NGC 2993 moderately bright, very small, round, weak concentration. A 13.5 mag star is 2′ SSE (SG 13″)
NGC 4038 NGC 4039	Appears as two irregular galaxies connected at the E end (SG 13″) Fairly bright, moderately large. Forms a striking "shrimp-like" or "comma" shape with the tail attached at the E end and extending to the S. Appears clearly darker between the two objects on the W side (SG 17.5″) Two long, very faint tidal tails can be traced out over 10′ from the roughly "u-shaped" pair of colliding galaxies. Numerous bright knots and clumps are visible – esp. in the larger galaxy – NGC 4038 (RJ 24″)
NGC 4085 NGC 4088	NGC 4085 is faint, small, elongated WSW–ENE. Two 8 mag stars are in the field to the SE and SW (SG 8″). Fairly faint, moderately large, very elongated 4:1 WSW–ENE, 2.5′×0.6′, weak concentration (SG 17.5″) NGC 4088 is fairly bright, elongated SW–NE, weak concentration, cigar-shaped (SG 8″). Bright, fairly large, elongated 5:2 SW–NE, 5.0′×2.0′, mottled patchy appearance, small elongated brighter core but no nucleus. A 15 mag star is 2′ off the NW side. Faint spiral structure visible with concentration. An extremely faint arm is off the NE end curving toward a 14.5 mag star to the NE 3.7′ from center and a second extremely faint arm is just visible off the SW end curving to the S (SG 17.5″)
NGC 4298 NGC 4302	NGC 4298 is fairly faint, slightly elongated NW–SE. A 13 mag star is at the E end (SG 13″). Fairly bright, moderately large, elongated NW–SE, broadly brighter center. A 13 mag star is at the E end 0.8′ from center. Forms a close pair with edge-on NGC 4302 2′ E (SG 17.5″) NGC 4302 is a faint edge-on streak N–S close following NGC 4298 (SG 13″). Fairly faint, large edge-on 7:1 N–S, 4.5′×0.6′, low surface brightness, weak concentration. A 14 mag star is off the N edge 2.0′ from center (SG 17.5″)
NGC 4567 NGC 4568	NGC 4567 is fairly faint, elongated E–W. NGC 4568 is attached at the NE end (SG 13″). Moderately bright, fairly small, elongated 3:2 ~E–W. Slightly smaller than NGC 4568 attached at the E end but has a slightly higher surface brightness (SG 17.5″)

(*Continued*)

Table 9.3. Visual descriptions of galaxy pairs—Cont'd

Pair	Description
NGC 4676A NGC 4676B	NGC 4676A is faint, small, low surface brightness, elongated N–S. In contact with NGC 4676B at the SE end. This object is the brightest of the pair and appears faint, small, round with a small bright core. The thin "tails" visible on photos not seen (SG 17.5")
NGC 5544 NGC 5545	Contact pair appearing as two brighter knots at the SW end (NGC 5544) and the NE end (NGC 5545). Separated by just a small darker region of lower surface brightness but are not cleanly resolved. NGC 5544 is a very elongated streak WSW–ENE, moderately large, uneven surface brightness NGC 5545 appears larger (SG 17.5")
ESO 138-29 ESO 138-30	ESO 138-29 is a faint 2′×1′ haze elongated NE–SW and broadly brighter on the SW end, with a small 30 in. diameter core, which brightens to the centre. There are suggestions of a darker patch E of the core and a faint ring structure with averted vision. Two 12 mag stars are off the SW end. ESO 138-30 is fairly faint 1′×0.2′ haze elongated NW–SE between two 12 mag stars. Averted vision shows an extremely faint bridge running SE–NW and connecting the cores of the two galaxies. Best viewed at 350× (Michael Kerr 25")
UGC 12914 UGC 12915	A close pair of elongated galaxies in roughly the same orientation. The brighter member UGC 12914 appeared as a faint, moderately large edge-on 4:1 NNW–SSE, 2.0′×0.5′. A 12.5–13 mag star is at the SE tip 1.6′ from the center. UGC 12915 lies just 1′ NE and is elongated 3:1 NW–SE, 1.2′×0.4′, with an even surface brightness (SG 17.5")

Additional Notes

NGC 2207/IC 2163. This beautiful pair of spiral galaxies is considered an interesting mix of overlapping and dynamically interacting systems. NGC 2207 is the closer object, as the bright knots and dusty motes of the large spiral arms can be seen cutting in front of the smaller IC galaxy.

NGC 2992/3. This is a pair of colliding spirals caught early in the "merging" process. Both galaxies show considerably activity in their cores, and NGC 2992 is a Seyfert galaxy whose core is heavily obscured by a heavy dust lane. Observers using large telescopes may glimpse long tidal tails and plumes.

NGC 3314. Perhaps the most dramatic of all overlapping galaxies, where a foreground face-on "pinwheel" spiral galaxy (NGC 3314a) is in the direct line of sight with a larger, inclined background spiral (NGC 3314b). The HST captured a superb image of this amazing pair recently (Fig. 2.5).

NGC 4038/9. "The Antennae" or "Ringtail Galaxies" (Fig. 1.35), this is one of the most intensely members of its class. Two large spirals are locked in a tightening cosmic dance, and huge tidal tails up to 15′ long have been thrown out [169]. Deep images have revealed hundreds of bright blue clusters, SSCs, and HII regions spawned by the collision. Over the next billion years or so, these galaxies will merge to form a new giant elliptical.

Table 9.4. Superimposed or overlapping galaxies

Object	Con	R.A.	Decl	V	V'	$a \times b$	PA	Type	Remarks
NGC 450	Cet	01 15 31.1	−00 51 36	11.8	13.8	3.1×2.3	70	SAB(s)cd	Very discordant redshifts
UGC 807		01 15 35.1	−00 50 52	15.0	14.0	0.9×0.4	38	Sc	
NGC 1738	Lep	05 01 46.6	−18 09 23	12.9	12.7	1.3×0.7	44	SB(s)bc pec	
NGC 1739		05 01 47.4	−18 10 00	13.5	13.3	1.4×0.7	105	SB(s)bc pec	
NGC 2207	CMa	06 16 21.8	−21 22 22	11.0	13.2	3.9×2.2	141	SBbc	
IC 2163		06 16 28.0	−21 22 35	11.7	12.9	3.0×1.2	98	Sbc	
NGC 3314	Hya	10 27 12.8	−27 41 00	13.1	13.0	1.5×0.7	143	Sab: sp	Spirals directly superimposed
NGC 4567	Vir	12 36 32.7	+11 15 28	11.3	13.2	3.1×2.2	85	Sbc	VV 219, Siamese Twins
NGC 4568		12 36 34.2	+11 14 19	10.9	13.3	4.6×2.2	23	Sbc	
M 60	Vir	12 43 40.3	+11 32 58	8.8	13.1	7.6×6.2	105	E2	Arp 116
NGC 4647		12 43 32.4	+11 34 56	11.3	13.2	2.9×2.3	102	SAB(rs)c	

Table 9.5. Visual descriptions of galaxy trios

Trio	Description
NGC 168 NGC 172 NGC 177	NGC 168 is very faint, very small, slightly elongated. An extremely faint star is possibly involved. NGC 172 is faint, edge-on 5:1 SSW–NNE, low even surface brightness. NGC 177 is brightest of the three. Faint, edge-on 4:1 N–S, bright core, stellar nucleus (SG 17.5″)
NGC 467 NGC 470 NGC 474	NGC 467 is fairly faint, small, round, weak concentration. NGC 470 is fairly faint, moderately large, diffuse, elongated 3:2 NNW–SSE, weak concentration at center. Largest of the three. NGC 474 is fairly bright, small, round, small bright core (SG 13″)
NGC 1618 NGC 1622 NGC 1625	NGC 1618 is faint, fairly small, very elongated 3:1 SSW–NNE, weak concentration. NGC 1622 is faint, elongated SW–NE, small bright core, stellar nucleus, faint elongated halo. NGC 1625 is fairly faint, edge-on 4:1 NW–SE, 1.4′×0.3′. A 14 mag star is at the NW tip 0.7′ from center. Located 10′ ENE of γ Eri (SG 17.5″)
NGC 2292 NGC 2293 NGC 2295	NGC 2292 is very faint, very small, round, low even surface brightness. NGC 2293 is fairly faint, small, round, very bright core stellar nucleus. NGC 2295 is faint, fairly small, very elongated 3:1 SSW–NNE, even surface brightness. Located between two 13 mag stars 30 in. SSW of center and 20 in. NNE or center (SG 17.5″)
M 105 NGC 3384 NGC 3389	M 105 is bright, very small bright core, slightly elongated. NGC 3384 is bright, bright stellar nucleus, elongated 5:2 SW–NE. NGC 3389 is fairly faint, very elongated 3:1 WNW–ESE, diffuse (SG 13″)
NGC 3991 NGC 3994 NGC 3995	NGC 3991 is moderately bright, fairly small, edge-on SSW–NNE, 1.0′×0.3′. This object has a bright stellar knot at the NNE end (about 25 in. from the center) giving an unusual asymmetric appearance! NGC 3994 is moderately bright, small, elongated 2:1 SSW–NNE, prominent core. NGC 3995 is moderately bright, moderately large, elongated 5:2 SW–NE, large bright core. Third and largest of an excellent trio (SG 17.5″)
NGC 4206 NGC 4216 NGC 4222	NGC 4206 is fairly faint, edge-on 6:1 exactly N–S, 4′×0.7′, fairly large, weak concentration. A 12 mag star lies 2.9′ SE of center. NGC 4216 is very bright, very large, edge-on 5:1 SSW–NNE, small very bright core. A 14 mag star is close E of the core. This is a striking galaxy and is the second of three edge-on galaxies. NGC 4222 is faint, moderately large, edge-on SW–NE, very thin. A 15 mag star is at the E end (SG 17.5″)
ESO 221-12 ESO 221-13 ESO 221-14	ESO 221-14 displays a soft, transparent haziness. With averted vision, the view improves to reveal a round shape and visible just north of an imaginary 10 mag double which is fairly obvious. Close and lengthy observation is required to see the faint shows of light. Even though I could identify the stars in the star field, I could not confirm the other two components (ESO 221-12/13). Toward the south just outside the galaxy two stars protrude against the background galaxy (Magda Streicher 12″)

NGC 6769 NGC 6770 NGC 6771	Quite interesting. The brightest of the three galaxies is NGC 6769 which appears as a quite faint 0.7′×0.5′ haze elongated SE–NW and with a slight brightening to the centre. The next brightest is NGC 6770 which appears as a round 0.3′ diameter haze with a slight brightening to the centre and an extension to the NW. NGC 6771 is a very faint, rectangular-shaped 0.4′×0.2′ haze elongated SE–NW with the impression of fainter extensions with averted vision (Michael Kerr 8″) Set in a rich field, NGC 6769 is a diffuse, low surface brightness 1.5′×1′ mottled disk elongated NW–SE with a slightly brighter 0.7′×0.5′ core and stellar nucleus. A 12 mag star is 40″ E and a 13 mag star is 1′ SE. NGC 6770 is a fairly faint, round 0.3′ diameter haze, which rises slightly to a stellar nucleus, and there are faint extensions NW and SE. It is surrounded by a very faint, diffuse, indistinct halo, and there is a slightly brighter knot at the end of the NW extension. The distorted spiral arm heading SW is not seen. Two 12/13 mag stars are 0.5′ E and 1′ ENE and a 12 mag star 1.5′ NNE. NGC 6771 is a fairly faint 1′×0.3′ haze elongated NW–SE with a brighter core. Best viewed at 350× (Michael Kerr 25″)
NGC 7232 NGC 7232B NGC 7233	NGC 7232 appears as soft, faint, and elongated thin dust lane, with a bright centre. In comparison NGC 7233 is slightly round and fainter, just east of NGC 7232. To the east two lovely orange stars round off the picture. The very faint NGC 7232B was not visible (Magda Streicher 8″). Interesting compact field of two galaxies and two bright stars. Brightest galaxy is NGC 7232, a moderately bright 2′×0.6′ haze, elongated E–W, which rises slightly to a central bulge, small core and stellar nucleus. NGC 7233 is low surface brightness 1′×0.8′ halo, elongated E–W, which rises abruptly to a small core elongated NE–SW and stellar nucleus. Two 9 mag stars are 1.6′ E and 1.9′ N. NGC 7232B is a very faint amorphous glow about 1′ in diameter best seen with averted vision, which rises to a central bar elongated NNW–SSE. Best viewed at 350× (Michael Kerr 25″)

NGC 4567/8. The "Siamese Twins," a beautiful duet of spiral galaxies whose true nature has been a subject of debate for decades. Though both galaxies are members of the Virgo Cluster, they are really a "line of sight" pairing rather than a true physical pair. Recent studies now indicate that NGC 4568 is the closer object [214].

NGC 4676 A/B. Also know as the "Mice" for this pair's unusual shapes and long tidal tails (Fig. 8.15). A much more distant and somewhat more evolved version of the "Antennae," they have been in a close gravitational dance for the past several 100 million years [169].

UGC 12914/5. A pair of interacting galaxies (Fig. 9.2) that is part of the Perseus Supercluster which lie about 60 Mpc distant. Between this pair is a well-developed tidal bridge that consists of luminous gaseous filaments, stretched out and ionized by the galaxies magnetic fields. Also around the larger galaxy (UGC 12914), there is a well-defined ring structure.

Fig. 9.1. II Zw 99 in Aquarius, an equilateral triangle of compacts

Fig. 9.2. The "taffy" galaxies UGC 12914/15 in Pegasus

Small Groups, Chains

For galaxy groups with four to eight members there is a celebrated (professional) source: Paul Hickson's *Atlas of Compact Groups of Galaxies* [94] with a listing of 100 galaxy groups. Hickson groups have long been popular targets for visual observers [170,171]. The brightest group is HCG 44, or the "Leo Quartet," featuring NGC 3185, NGC 3187, NGC 3190, and NGC 3193. Other prominent examples are "Stephan's Quintet" [172], "Seyfert's Sextet" (Fig. 9.3), and "Copeland's Septet" (Fig. 3.9). Despite some bright examples, Hickson groups are generally difficult visual targets. Even more challenging are Shakhbazian groups, which are defined as "compact groups of compact galaxies." The brightest example is Shkh 30 (HCG 97).

Stephan's Quintet is situated about 34′ southwest of the bright spiral NGC 7331 (Fig. 7.10), which is located in the "middle" of a group of background galaxies [173]. Also in the constellation of Pegasus we find the bright NGC 7385 group [174]. Another interesting collection is the IC 2199 group, located only 30′ south of Castor. Years ago, Walter Scott Houston first mentioned this in his Sky & Telescope column *Deep Sky Wonders*. It is not the only remarkable target in Gemini [175]. Another nice target is the NGC 128 group [245]. The following Table 9.6 lists some examples of small groups of galaxies. For data of their members, consult the sky atlas software or common databases.

We will now discuss a special type of small groups showing a linear alignment of their members: galaxy chains [176]. One particularly well-known example is

Fig. 9.3. Seyfert's Sextet (VV 115) in Serpens

Table 9.6. Groups of 4–7 galaxies. Position is for center; *N* = number of galaxies; *d* = group diameter (arcmin); *V* = visual magnitude range

Group	Name	Con	R.A.	Decl	*N*	*d*	*V*	Galaxies
Phoenix group	Rose 34	Phe	00 21.4	–48 38	4	8	12.9–14.1	NGC 87, 88, 89, 92
NGC 128 group		Psc	00 29.1	+02 51	5	9	11.8–14.8	NGC 125, 126, 127, 128, 130
IC 2199 group		Gem	07 33.9	+31 24	5	30	12.8–15.4	IC 2192, 2193, 2194, 2196, 2197, 2199
VV 116	HCG 40, Arp 321	Hya	09 38.9	–04 51	5	3	12.8–16.0	MCG -1-25-8 thru -12
Leo Quartet	HCG 44, Arp 316	Leo	10 18.0	+21 49	4	20	10.8–12.9	NGC 3185, 3187, 3190, 3193
Copeland's Septet	HCG 57, Arp 320	Leo	11 37.8	+21 59	7	6.5	13.7–15.2	NGC 3745, 3746, 3748, 3750, 3751, 3753, 3754
The Box	HCG 61	Com	12 12.4	+29 11	4	8	12.2–13.5	NGC 4169, 4173, 4174, 4175
NGC 5044 group		Vir	13 15.4	–16 27	7	20	10.8–14.2	NGC 5035, 5037, 5044, 5046, 5047, 5049, MCG -3-34-33
NGC 5353 group	HCG 68	CVn	13 53.5	+40 19	5	13	11.1–13.7	NGC 5350, 5353, 5354, 5355, 5358
NGC 5629 group	AMW 3	Boo	14 28.2	+25 52	5	10	12.1–14.7	NGC 5629, IC 1013, 1017, 1018, 1019
Seyfert's Sextet	VV 115	Ser	15 59.2	+20 45	5	2.5	13.2–15.4	NGC 6027A-E
Pavo group	VV 297	Pav	20 18.1	–70 50	7	17	10.7–14.6	NGC 6872, 6876, 6877, 6880, IC 4970, 4972, 4981
Stephan's Quintet	HCG 92, Arp 319	Peg	22 36.0	+33 58	5	6	12.5–13.6	NGC 7317, 7318 A/B, 7319, 7320
NGC 7385 group		Peg	22 49.9	+11 37	6	16	12.2–14.3	NGC 7383, 7385, 7386, 7387, 7389, 7390
Shkh 30	HCG 97	Psc	23 47.4	–02 18	5	7	13.0–15.5	IC 5351, 5356, 5357, 5359, PGC 72405

Fig. 9.4. The master's "necklace": 8 Zw 388 in Virgo, a unique object

Markarian's Chain, located near the heart of the Virgo Cluster. Easily visible in even a small scope, it has a good variety of different Hubble galaxy classes. Stretching over two degrees from M 84 to NGC 4477 it includes nine large galaxies and many dwarf systems. Even stranger are "rings of galaxies," but there is only one example: Zwicky's "necklace" 8 Zw 388. This "ring" is extremely faint, but maybe a suitable target for CCD imaging with telescopes of sufficient focal length (Fig. 9.4). Table 9.7 presents a collection of chains, including the "necklace" (visual descriptions in Table 9.8). Figure 9.5 presents a nice chain of 5 MCG galaxies near M 51.

Clusters

Before listing Abell rich clusters, we will examine some medium-sized, less concentrated aggregates (not in Abell's catalogue), plus larger groups and poor clusters (Fig. 9.6). Prominent examples are the groups around IC 698 [177], NGC 383 [178], NGC 507 [179], or NGC 4005 [180,251], and the Pegasus I Cluster [181,182,248]. These and others are presented in Table 9.9 (observations in Table 9.10). The Virgo Cluster (which bears no Abell number) will not be discussed here because of it's overly large size and numerous galaxies (see Webb Society Handbook Vol. 5 or Luginbuhl & Skiff).

Table 9.7. Galaxy chains and a galaxy ring

Group	Con	A.R.	Decl	*N*	*d*	*V*	Galaxies	Remarks
Burbidge Chain	Cet	00 47.6	–20 28	4	6	14.5–16.5	MCG -4-31-10 thru -13	VV 518, 20′ n of NGC 247
VV 172	Dra	11 32.1	+70 49	5	1	15.2–16.8	UGC 6514 a-e	Arp 329, HCG 55
VV 150	UMa	11 32.6	+52 57	5	3	14.5–16.5	UGC 6527 a-e	Arp 322, HCG 56, 8′ s of NGC 3718
MKW 3	Vir	11 49.6	–03 31	5	8	14.5–15.8	CGCG 12-93, -97, -98, -99, and anon	
Markarian's Chain	Vir	12 28.0	+13 10	9	180	8.9–11.8	M 84, 86, NGC 4435, 4438, 4458, 4461, 4473, 4477, 4479	Virgo Cluster
Centaurus Chain	Cen	12 44.3	–40 44	4	12	11.8–14.7	NGC 4622, 4622B, 4650, 4650A, PGC 42911	Centaurus Cluster
HCG 66	UMa	13 38.6	+57 19	4	1.3	15.2–17.0	PGC 48220, 48222, 48226, 48231	Not VV 135
MCG chain	CVn	13 18.5	+47 12	5	8	15.8–17.2	MCG 8-24-102 thru -105, UGC 8364	2° W of M 51
Necklace	Vir	14 15.2	–00 29	10	2	18.0–20.0	8Zw 388	
Shkh 166	UMi	16 52.2	+81 37	11	15	14.5–16.8	MCG 14-8-15 thru -18, PGC 59120, 59211, NPM1G +81.91, and 4 anon	UGC 10638, A 2247
Klemola 30	Tel	20 00.9	–47 05	5	2	13.0–15.4	NGC 6845, 6845A-D	
HCG 88	Aqr	20 52.5	–05 44	4	7	13.3–14.0	NGC 6975-78	

Table 9.8. Visual descriptions of galaxy groups and chains (omitting "Markarian's Chain;" for M 84 and M 86 see Table 7.2)

Group	Description
NGC 128 group	NGC 125 fairly faint, small, round, very bright core, stellar nucleus. NGC 126 very faint, very small, oval ~E–W? NGC 127 very faint, very small, round. NGC 128 moderately bright, fairly small, very elongated 3:1 N–S, bright core, stellar nucleus. Brightest in a group of five with two extremely close companions: NGC 127 0.8′ NW and NGC 130 1.0′ ENE. NGC 130 very faint, very small, oval ~SW–NE, small bright core (SG 17.5″)
NGC 5044 group	NGC 5035 fairly faint, fairly small, slightly elongated N–S, weak concentration. NGC 5037 moderately bright, fairly small, elongated 2:1 SW–NE, bright core. A 13.5 mag star is at the NE tip. NGC 5044 fairly bright, round, 2.0′ diameter, moderate concentration. NGC 5046 faint, very small, slightly elongated. NGC 5047 fairly faint, fairly small, edge-on 4:1 E–W, small bright core. NGC 5049 fairly faint, small, slightly elongated, small bright core (SG 17.5″)
NGC 5353 group	NGC 5350 fairly faint, diffuse, slightly elongated, very weak concentration, no core. NGC 5353 fairly bright, oval 2:1 NW–SE, gradually increases to a small bright core. Forms a close pair with NGC 5354 1.2′ N. NGC 5354 fairly faint, fairly small, broad concentration. NGC 5355 faint, small, slightly elongated SSW–NNE, even surface brightness. NGC 5358 very faint, very small, very elongated NW–SE (SG 17.5″)
NGC 7385 group	NGC 7383 faint, small, slightly elongated, brighter core. NGC 7385 moderately bright, broadly concentrated halo, small bright core, slightly elongated ~N–S. A 11 mag star is 1.0′ NW. NGC 7386 fairly faint, slightly elongated NNW–SSE, very small bright core. NGC 7387 faint, very small, slightly elongated, gradually increases to a very small bright core. NGC 7389 faint, very small, brighter core, slightly elongated. Forms a close pair with NGC 7390 2.3′ SSE. NGC 7390 very faint, very small, round, low even surface brightness (SG 17.5″)
IC 2199 group	IC 2193 faint, small, elongated 5:2 WSW–ENE, small bright core. A 13.5 mag star is at the N edge 32 in. NNE of center. IC 2194 faint, very small, round, very small bright core. IC 2196 fairly faint, fairly small, round, even concentration to a brighter core. IC 2199 fairly faint, fairly small, elongated 2:1 SW–NE, brighter along major axis (SG 17.5″)
HCG 88	NGC 6976 extremely faint, very small, round, very diffuse. NGC 6977 very faint, fairly small, round, diffuse, even surface brightness. NGC 6978 fairly faint, fairly small, bright core, elongated 2:1 NW–SE (SG 17.5″)
VV 116	MCG -1-25-9 very faint, very small, round. MCG -1-25-8 is an elongated threshold object glimpsed intermittently just N of MCG -1-25-10. At first, this pair was not resolved and I was not sure if I was viewing a single compact or elongated gx, but was gradually convinced that two distinct galaxies were visible. MCG -1-25-10 appears very faint, extremely small, round. At times, appears elongated or a fainter companion system is attached at the N side (this is MCG -1-25-8). MCG -1-25-12 appears very faint, small, elongated 3:2 E–W. This is the second easiest (of 4) in an interesting tight group. A 14 mag star lies 2.3′ NW (SG 17.5″)

(*Continued*)

Table 9.8. Visual descriptions of galaxy groups and chains (omitting "Markarian's Chain;" for M 84 and M 86 see Table 7.2)—Cont'd

Group	Description
VV 150	Very difficult. Three galaxies of chain only visible as elongated string, very faint. Edge-on galaxy E very faint, averted vision (20″)
VV 172	Appears as an extremely faint, elongated string SSW–NNE about 1′ in length. Faint enough to require averted vision but appears irregular. At 280×, a couple of individual components (A and either B or C) are sometimes resolved with the more obvious "knot" at the N end of the string (HCG 55a) appearing barely nonstellar. This well known chain contains a discordant redshift (55e) and is located 25′ NW of NGC 3735 (SG 17.5″, 220×)
Shkh 30	IC 5351 extremely faint, extremely small, round, 10 in. diameter. Attached at the N side of a 11 mag star which makes viewing very difficult. IC 5356 very faint, very small, slightly elongated (although difficult to pin down direction), very weak concentration. IC 5357 faint, small, elongated 3:2 NW–SE, 0.7′×0.4′, gradually brightens to a small bright core and an almost stellar nucleus. IC 5359 extremely faint, small, very elongated 4:1 NW–SE. Only visible with averted vision and cannot be held steadily. Located 1.6′ ENE of a 10 mag star which also detracts from viewing (SG 17.5″)
Shkh 166	MCG 14-8-17 faint, round, about 20 in., very weak concentration. MCG 14-8-15 very faint and small, round, 15–20 in. diameter. Requires averted for best view. Located less than 2′ NW of a 12.5 mag star. MCG 14-8-16 very faint, round, ~20 in. diameter. MCG 14-8-18 extremely faint, very small, requires averted to comfortably view. In moments of steady seeing the galaxy is clearly elongated 2:1 ~N–S with dimensions ~20 in.×10 in. NPM1G +81.0091 very faint, round, ~15 in. diameter. Situated just 0.7′ SW of a 13 mag star (SG 17.5″)
Klemola 30	Very difficult. NGC 6845A can be just seen with direct vision as a very faint 0.7′×0.5′ haze elongated NE–SW. NGC 6845B is visible most of the time with averted vision as a very small, extremely faint hazy spot. NGC 6845C can be glimpsed occasionally as an extremely faint haze close to NGC 6845A. NGC 6845D is not seen. Best viewed at 230× (Michael Kerr 8″) Nice group. Best viewed at 350× or 450×. All galaxies are seen with direct vision. NGC 6845A is the brightest appearing as a 1′×0.5′ haze elongated NE–SW with suggestions of structure. NGC 6845C is a thin bright 0.7′×0.2′ spindle. NGC 6845B is an amorphous 0.4′ diameter haze. NGC 6845D is a small faint 0.2′×0.1′ haze (Michael Kerr 25″)
Burbidge Chain	MCG -4-3-10 is the brightest and furthest north. At 200× appeared faint, small elongated 5:2 SSW–NNE, 0.8′×0.3′, with an even surface brightness. A 12 mag star is 1.2′ north. MCG -4-3-11 is a marginal galaxy in the chain and is sandwiched between MCG -4-3-10 just 3.4′ north and MCG -4-3-13 2.0′ S. I required averted vision in 6.0 mag skies and only popped into view momentarily as a threshold 15″ knot. A 14 mag star lies 1.5′ W. The further south of the trio is MCG -4-3-13 which appeared extremely faint, small, roundish ~25 in. in diameter. This galaxy also required averted vision though it could be almost continuously held with concentration (SG 17.5″)

Centaurus Chain	NGC 4622 very faint, small, round, low fairly even surface brightness. NGC 4622B very faint, very small, round. NGC 4650 very faint, small, oval WNW–ESE, bright core. NGC 4650A not recorded (SG 17.5″)
Copeland's Septet	NGC 3745 extremely faint and small, round. NGC 3746 very faint, very small, round. NGC 3748 extremely faint, extremely small, round. NGC 3750 faint, very small, round, very small bright core. NGC 3751 extremely faint, extremely small, round, 20 in. diameter. Requires averted vision although easier to view than NGC 3754. NGC 3753 very faint, very small, slightly elongated NW–SE. NGC 3754 extremely faint and small, round. Difficult to resolve from brighter NGC 3753 just 40 in. SW of center. A 12 mag star is 1.0′ N (SG 17.5″)
Leo Quartet	NGC 3185 fairly faint, gradually brighter core. NGC 3187 very faint, elongated NW–SE. NGC 3190 bright, small bright nucleus, elongated NW–SE. NGC 3193 bright, small bright nucleus, small, round. A 9 mag star is just 1′ N (SG 13″)
MCG chain	All five galaxies detected. UGC 8364 at south end, edge-on, most difficult, star near (Frank Richardsen 20″)
Pavo group	NGC 6872 moderately bright, fairly small, elongated 2:1 SW–NE in the direction of a 10.4 mag star 1.1′ WSW of center, ~1.2′ 0.6′, broad concentration with a brighter core. Interacting with IC 4970 at 1.1′ N just outside the halo. NGC 6876 moderately bright and large, slightly elongated ~E–W, 1.5′×1.3′, containing a brighter core. A star is at the south edge 0.5′ from center. Forms a close pair with NGC 6877 just 1.5′ following. NGC 6877 faint, very small, oval N–S, 0.3′×0.15′. NGC 6880 faint, small, elongated 5:2 SSW–NNE, 0.5′×0.2′. A 13 mag star is at the west edge. Forms a close pair with IC 4981 off the NE edge 1.1′ from the center. IC 4970 faint, very small, slightly elongated, 20 in.× 15 in. A 10.4 mag star is 1.8′ SW. IC 4972 appears with averted vision as an extremely faint, ghostly streak was just visible oriented SSW–NNE, ~0.5′×0.1′ with a low, even surface brightness. IC 4981 very faint, very small, 20 in. diameter (SG 18″, 171×)
Phoenix group	NGC 92 is quite faint but seen with averted vision as a 0.7′×0.5′ haze elongated NW–SE. NGC 89 is only seen with averted vision S of a 14.5 mag star as a very faint 0.7′×0.3′ haze elongated NW–SE. NGC 87 is a very faint 0.5′ diameter round haze visible half the time with averted vision. NGC 88 is an extremely faint 0.3′ diameter round haze occasionally glimpsed with averted vision. Best viewed at 230× (Michael Kerr 8″) Nice grouping of galaxies, all visible with direct vision. The brightest is NGC 92, a 1′×0.7′ haze elongated NW–SE with no obvious core. The edges of the halo are diffuse and there are suggestions of a NW extension and a longer one to the SE. Next brightest is NGC 89, a 1′×0.5′ haze elongated NW–SE, which is broadly brighter to the centre. Averted shows faint extensions hooking away at either end. NGC 87 is a 0.8′ round haze with diffuse even surface brightness and very little central brightening. There is a suggestion of a brighter nucleus with averted vision. NGC 88 is a 0.5′ diameter round haze with a stellar core and a faint star close SW. Best viewed at 353× (Michael Kerr 25″)

(*Continued*)

Table 9.8. Visual descriptions of galaxy groups and chains (omitting "Markarian's Chain;" for M 84 and M 86 see Table 7.2)—Cont'd	
Group	Description
Seyfert's Sextet	On close inspection, the confused "clump" resolves into three components with the brightest component (NGC 6027E) appearing faint, small, elongated ~E–W. Extremely close by are NGC 6027A just 36 in. SSW and NGC 6027B 22 in. W of center. A 14.5 mag star is 1.1′ ESE and other faint stars are near. These three galaxies are just resolved at 220× (SG 17.5″)
Stephan's Quintet	NGC 7317 very faint, small, round. A star is at the NW edge. NGC 7318 faint, elongated, two stellar nuclei visible in good seeing. NGC 7319 extremely faint, fairly small, requires averted vision (SG 13″). NGC 7320 extremely faint, small (SG 8″). Moderately bright, moderately large, brighter core, elongated 5:2 NW–SE. A 14.5 mag star is at the SE side 15 in. from the center (SG 17.5″)
The Box	NGC 4169 moderately bright, fairly small, slightly elongated NNW–SSE, very small bright core. NGC 4173 very faint, very elongated NW–SE, low even surface brightness. NGC 4174 fairly faint, prominent very small bright core. Slightly elongated halo is faint and small. NGC 4175 faint, edge-on NW–SE, bright core, similar in size to NGC 4173 but fainter (SG 13″)

The Abell clusters are those presented here also include the southern extension of the catalogue (by Abell, Olowin, and Corwin). Visually the selection, presented in Table 9.11, ranges from "easy," like A 194 [249], A 262 [183], or A 1367 [251] to "extremely difficult," as in the case of the celebrated Corona Borealis Cluster (A 2065) [184]. Be aware that "observation" only means that a few of the brightest members may be detected. In the case of A 2065, there will be never more than six galaxies visible with 18–20″ aperture, even under the best conditions. Figure 9.7 shows the famous "Haufen A" (A 151) in Cetus.

Many rich clusters are constituents of superclusters. Thus you can reach the highest step of cosmic hierarchy by observing targets like A 426, A 2151, or A 1656, which are the dominant members in the Perseus-Pisces [185], Hercules [186], and Coma/A 1367 superclusters [187]. Further literature on visual observations of individual clusters: A 119 [188], A 347 [189], A 1314 [190], A 2197/99 [191], A 3526 [192], a member of the Hydra-Centaurus Supercluster [193], A 4038 [194].

Additional Notes

A 194. This is a fairly close galaxy cluster that is part of the Perseus-Pisces Supercluster. Its members include the giant galaxy NGC 541 (Type S0) and the dumbbell galaxy pair, NGC 545/7 [185]. In the NE halo of NGC 541 is "Minkowski's Object" (see Table 10.4) – an extragalactic H II region/galaxy fragment. This is an extremely difficult and tiny object and requires the largest sized apertures for any chance of success.

Fig. 9.5. Close to M 51 in Canes Venatici, but pretty unknown, the "MCG chain" of five faint fuzzys

A 347. A cluster that is a member of the Perseus-Pisces Supercluster, located on the western outskirts of A 426. Though not as dense as many of the other Abell clusters on this list, its main claim to fame is its close proximity to the beautiful edge-on system NGC 891 (Fig. 6.6).

A 426. The Perseus Cluster (Fig. 2.11) is the closest "rich cluster" with richness class 2, according to the Abell's classification scheme. The most prominent member is the giant cD galaxy NGC 1275, though there are a number of other giant ellipticals near the core. Many of the spiral galaxies are "early types," or "anemic," that is, having much of their gas and dust stripped off either by close encounters or by "ram pressure" via orbiting in the dense extragalactic medium.

A 779. This cluster is located less than a degree to the SW of the bright star α Lyn. Over 700 million ly distant, many of its members are quite faint. By far the most impressive is the 12 mag cD galaxy NGC 2832, located in the heart of the cluster. Several other much fainter systems including NGC 2831 and NGC 2830 are located within the galaxy's extensive halo.

A 1060. The Hydra I Cluster is a fairly irregular cluster of richness class 1. Its core is dominated by a pair of cD-like galaxies (NGC 3309 and NGC 3311), and its

Fig. 9.6. The poor cluster AMW 7 in Perseus with cD galaxy NGC 1129

membership includes the overlapping systems NGC 3314 (see Table 9.4). It is part of the Hydra-Centaurus Supercluster complex.

A 1185. This is a fairly rich cluster (richness class 1) for the visual observer that includes six NGC galaxies. The brightest, NGC 3550, is part of a small chain of galaxies in the western half of the cluster. "Ambartsumian's Knot" or NGC 3561A is a small, peculiar elliptical galaxy (Arp 105) located in the eastern half of the cluster.

A 1367. Visually this is a very rich cluster in medium to large sized telescopes. Over 60 galaxies are visible in an area of a square degree, and it contains more objects brighter than 14 mag than either the Coma Cluster (A 1656) or Hercules Cluster (A 2151). The giant elliptical (NGC 3842) is surrounded by a swarm of smaller spirals and ellipticals, while other smaller knots are distributed haphazardly around the cluster.

A 1656. The Coma Cluster (Fig. 1.2) is one of the best known of all Abell clusters. Located at a distance of over 300 million ly, its two huge cD galaxies – NGC 4874 and NGC 4889 – dominanting its core region. This cluster, plus A 1367, A 2151, and others, form what has become known as the "Great Wall": a massive sheet-like structure that spans hundreds of millions of ly. This cluster is a visual treat in larger telescopes, as hundreds of galaxies are visible over a span of several square degrees. Like A 426,

Table 9.9. Sample of small (poor) clusters with its associated NGC/IC galaxies. Position is for center; N = number of galaxies; d = group diameter (arcmin); V = magnitude range

Cluster	Con	A.R.	Decl	N	d	V	NGC/IC Galaxies
Cancer Cluster (center)	Cnc	08 20.9	+21 04	20	40	12.3–15.5	NGC 2556, 2560, 2562, 2563, 2569, 2570
NGC 383 group	Psc	01 07.4	+32 24	14	20	12.2–16.5	NGC 373, 375, 379, 380, 382–388
NGC 507 group	Psc	01 33.4	+33 20	20	30	11.3–16.0	NGC 494, 495, 496, 498, 501, 503, 504, 507, 508, IC 1684, 1685, 1688, 1689, 1690
NGC 4005 group	Leo	11 58.1	+25 09	13	30	12.3–14.9	NGC 3987, 3989, 3993, 3997, 3999, 4000, 4005, 4011, 4015, 4018, 4021, 4022, 4023
Pegasus I Cluster (center)	Peg	23 20.0	+08 16	20	35	11.1–16.0	NGC 7608, 7611, 7615, 7616, 7617, 7619, 7621, 7623, 7626, IC 5309
MKW 4	Vir	12 04.2	+01 51	13	15	11.4–16 5	NGC 4063, 4073, 4077, 4139 (= IC 2989)
AMW 7	Per	02 54.6	+41 35	10	12	11.9–16.5	NGC 1129, 1130, 1131, IC 265
NGC 5416 group	Boo	14 02.9	+09 30	15	30	13.0–15.5	NGC 5409, 5416, 5423, 5424, 5431, 5434, 5436–38

Table 9.10. Visual descriptions of small clusters

Cluster	Description
Cancer Cluster (center)	NGC 2556 very faint, very small, round. NGC 2560 faint, small, very elongated 3:1 E–W, small bright core. NGC 2562 fairly faint, small, oval 3:2 N–S, halo brightens to a small bright core. NGC 2563 fairly faint, fairly small, almost round, halo brightens evenly to a small bright core. Appears similar to NGC 2562 4.7′ NW but slightly larger. NGC 2569 very faint, very small, round, small bright core in low surface brightness halo. NGC 2570 very faint, small, very low surface brightness. Slightly larger than NGC 2569 2.6′ S but has a lower surface brightness (SG 17.5″)
NGC 383 group	NGC 373 very faint, very small, slightly elongated ~E–W. NGC 375 extremely faint, extremely small, round. NGC 379 fairly faint, fairly small, elongated ~N–S, even surface brightness. NGC 380 fairly faint, small, round, bright core, stellar nucleus. NGC 382 faint, very small, round. Forms a double system with much brighter NGC 383 30″ NNE. NGC 383 brightest in the NGC 383 cluster. Fairly bright, moderately large, slightly elongated, broadly concentrated halo. NGC 384 fairly faint, slightly elongated, bright core. NGC 385 fairly faint, small, slightly elongated, bright core. NGC 386 very faint, very small, round, bright core. NGC 387 extremely faint, round, almost stellar. NGC 388 extremely faint, extremely small, round, size 10–15 in. (SG 17.5″)
NGC 507 group	NGC 494 fairly faint, very elongated 3:1 ~E–W, bright core. NGC 495 faint, small, slightly elongated, small bright core. NGC 496 faint, low even surface brightness. NGC 504 faint, small, very elongated 3:1 SW–NE, small bright core. NGC 507 moderately bright, moderately large, round, very bright core. NGC 508 fairly faint, small, round (SG 13″) NGC 498 extremely faint and small, no details visible. This very difficult object was only detected after extended viewing at 220×, 280×, and 420×. NGC 501 very faint, very small, round, 20 in. diameter. Can just hold continually with averted vision once identified. NGC 503 very faint, very small, round, 20 in. diameter. IC 1685 very difficult. Just glimpsed with averted vision at 280× and appeared as a 10 in. fleeting spot with no concentration. A 14.5 mag star lies 45″ SSE. IC 1690 extremely faint, very small, elongated 2:1 NW–SE, 20 in.×10 in. Extended in the direction of a 12 mag star 1.5′ SE (SG 17.5″)
NGC 4005 group	NGC 3987 fairly faint, moderately large, edge-on WSW–ENE, weak concentration. NGC 3989 extremely faint, very small, round. NGC 3993 faint, fairly small, very elongated NW–SE, broad concentration. Forms a pair with NGC 3989 2.7′ WSW. NGC 3997 faint, small, elongated ~E–W (central bar), small bright core. NGC 3999 extremely faint, very small, round, 15 in. diameter. Requires averted vision and can only hold steadily 2/3 of the time. NGC 4000 very faint, fairly small, very elongated N–S. NGC 4005 fairly faint, small, oval slightly elongated E–W, bright core. NGC 4015 fairly faint, small, slightly elongated bright core. NGC 4018 faint, fairly small, edge-on NW–SE. NGC 4021 very faint, very small, slightly elongated ~E–W, 0.4′×0.3′. NGC 4022 faint, small, slightly elongated, bright core. NGC 4023 faint, small, slightly elongated ~N–S, weak concentration (SG 17.5″)

Pegasus I Cluster (center)	NGC 7608 very faint, small, diffuse, very elongated ~N–S, even surface brightness, requires averted vision. NGC 7611 fairly faint, small, elongated 2:1 NW–SE, stellar nucleus. NGC 7615 very faint, diffuse, slightly elongated ~E–W. A 14 mag star is off the E edge 1.0′ from the center. NGC 7616 faint, small, slightly elongated oval, brighter core. NGC 7617 faint, small, elongated 3:2 N–S, 0.9′×0.6′, weak even concentration to a brighter core. NGC 7619 bright, elongated, bright core, stellar nucleus. NGC 7621 very faint, small, elongated, requires averted vision. NGC 7623 fairly bright, small, elongated, bright core, stellar nucleus, very faint extensions ~N–S. NGC 7626 bright, slightly elongated 4:3, brighter core (although less intense than NGC 7619), substellar nucleus. IC 5309 faint, very elongated SSW–NNE (SG 17.5″)
MKW 4	NGC 4063 very faint, very small, slightly elongated N–S. NGC 4073 moderately bright, elongated WNW–ESE, moderately large, bright core, stellar nucleus. NGC 4077 fairly faint, oval ~N–S. A 14 mag star is attached at the N end. NGC 4139 faint, very small, elongated 2:1 SW–NE, small bright core (SG 17.5″)
AMW 7	NGC 1129 moderately bright, moderately large, elongated WSW–ENE, brighter along major axis, small bright core. A 15 mag star is at the W edge 22 in. from the center. NGC 1130 very faint, very small, round. NGC 1131 very faint, very small, round, bright core. IC 265 not seen (SG 17.5″)
NGC 5416 group	NGC 5409 fairly faint, slightly elongated SW–NE, 1.2′×1.0′. Just a very weak even concentration to a slightly brighter core and an occasional faint stellar nucleus. Halo fades into background without a distinct edge. NGC 5416 moderately bright, elongated 3:2 WNW–ESE, 1.4′×0.9′, broad concentration. NGC 5423 fairly faint, small, round, 40 in. diameter, sharp concentration with a very small bright core and occasional stellar nucleus surrounded by a fainter halo. NGC 5424 fairly faint, round, 1.2′ diameter, small bright core. A 14 mag star is 1.0′ S. NGC 5431 faint, round, 0.6′ diameter, low surface brightness glow with no concentration. NGC 5434 is a close double system with the western component (A) larger and brighter. Fairly faint, slightly elongated SW–NE, 1.2′×1.0′, very little concentration. Contact pair with B at the NE end separation 1.5′. B is faint, very elongated 3:1 ~E–W, 1.0′×0.3′, low surface brightness, no concentration. NGC 5436 faint, very small, faint halo with an abrupt brighter core. NGC 5437 faint, small, round, even surface brightness. NGC 5438 faint, small, round, weak even concentration to a brighter core and occasional faint stellar nucleus (SG 17.5″)

Table 9.11. A selection of bright or nearby Abell clusters (sorted by Abell number). m_{10} = magnitude of 10th brightest member, Dist = distance in Mpc, N = number of galaxies, a = size (arcmin), Galaxy = dominant member, SC = supercluster membership. N8 and N16 counts the number of galaxies visible in an 8″ or 16″ telescope under good conditions (numbers in brackets denote that 18″ or more is needed)

A	Con	R.A.	Decl	z	Dist	M_{10}	N	a	Galaxy	Remarks	SC	N8	N16
71	And	00 37.8	+29 35	0.0724	296	15.5	30	19	NGC 183				6
76	Cet	00 39.8	+06 46	0.0405	168	15.0	42	28	IC 1565		Psc-Cet		4
119	Cet	00 56.4	−01 15	0.0442	183	15.0	69	39	UGC 579		Psc-Cet		4
151	Cet	01 08.9	−15 25	0.0533	220	15.0	72	39	IC 80A	Haufen A	Psc-Cet		5
194	Cet	01 25.6	−01 30	0.0180	75	13.9	37	56	NGC 541			5	12
262	And	01 52.8	+36 08	0.0163	68	13.3	40	100	NGC 708		Per-Psc	2	14
347	And	02 25.8	+41 52	0.0184	77	13.3	32	56	NGC 906		Per-Psc		7
426	Per	03 18.6	+41 30	0.0179	75	12.5	88	190	NGC 1275	Perseus	Per-Psc	4	17
569	Lyn	07 09.2	+48 37	0.0201	84	13.8	36	26	NGC 2329			2	4
634	Lyn	08 14.6	+58 02	0.1890	735	14.9	40	28	UGC 4280A				6
779	Lyn	09 19.8	+33 46	0.0225	94	13.8	32	50	NGC 2832			5	10
1060	Hya	10 36.9	−27 30	0.0126	53	12.7	50	168	NGC 3309	Hydra I	Hya-Cen	6	50
1185	UMa	11 10.8	+28 40	0.0325	135	14.3	52	28	NGC 3550		Leo		6
1213	UMa	11 16.5	+29 15	0.0469	194	14.5	51	22	UGC 6292				4
1228	UMa	11 21.5	+34 19	0.0352	146	13.8	50	50	UGC 6394		Leo		4
1314	UMa	11 34.8	+49 02	0.0335	139	13.9	44	28	IC 708		Leo		4
1367	Leo	11 44.5	+19 50	0.0220	92	13.5	117	100	NGC 3842	Leo Cluster	Com/A1367	2	50
1377	UMa	11 47.0	+55 44	0.0514	212	15.0	59	20	MCG 9-19-196	Ursa Major I			3
1656	Com	12 59.8	+27 58	0.0231	97	13.5	106	220	NGC 4874	Coma Berenices	Com/A1367	2	50
2065	CrB	15 22.7	+27 43	0.0726	296	15.6	109	22	MCG 5-36-20	Corona Borealis	CrB		(6)
2079	CrB	15 28.1	+28 52	0.0690	282	15.4	57	18	UGC 9861		CrB		3
2147	Her	16 02.3	+15 53	0.0350	145	13.8	52	39	UGC 10143		Her		7
2151	Her	16 05.2	+17 44	0.0366	152	13.8	87	56	NGC 6042	Hercules	Her		14
2162	CrB	16 12.5	+29 32	0.0322	134	13.7	37	56	NGC 6086		Her	2	4

2197	Her	16 28.2	+40 54	0.0308	128	13.9	73	90	NGC 6146		Her		6
2199	Her	16 28.6	+39 31	0.0302	126	13.9	88	90	NGC 6166		Her		5
2572	Peg	23 18.4	+18 44	0.0403	167	15.3	32	28	NGC 7571				4
2593	Peg	23 24.5	+14 38	0.0413	171	15.1	42	28	NGC 7649				4
2634	Peg	23 38.3	+27 01	0.0314	131	13.8	52	22	IC 5342				5
2666	Peg	23 50.9	+27 08	0.0268	112	13.8	34	78	NGC 7768			1	7
2877	Phe	01 09.8	–45 54	0.0247	103	14.3	30	28	IC 1633		Scl		4
3389	Dor	06 21.8	–64 57	0.0267	111	14.6	35	28	NGC 2235				4
3390	Col	06 25.0	–37 20	0.0333	138	14.7	63	28	ESO 365-16				1
3526	Cen	12 48.9	–41 18	0.0114	48	13.2	33	180	NGC 4696	Centaurus	Hya-Cen	3	20
3537	Cen	13 01.0	–32 26	0.0320	133	14.3	35	28	ESO 443-24		Hya-Cen		3
3565	Cen	13 36.7	–33 58	0.0123	52	14.0	64	56	IC 4296		Hya-Cen		9
3574	Cen	13 49.2	–30 17	0.0160	67	13.4	31	56	IC 4329		Pav-Ind		9
3627	Nor	16 15.5	–60 54	0.0157	66	13.5	59	56	ESO 137-8	Great Attractor	Pav-Ind		10
3656	Sgr	20 00.5	–38 31	0.0190	80	13.6	35	56	IC 4931				5
4038	Scl	23 47.7	–28 08	0.0300	125	14.2	117	28	IC 5358	Center = Klemola 44			10
S 373	For	03 38.5	–35 27	0.0046	19	10.3	50	180	NGC 1399	Fornax		12	27

Fig. 9.7. The rich cluster A 151 (Haufen A) in Cetus

there are numerous elliptical and early-type galaxies, especially in the dense core region. Most of the late-type spirals are located in the outskirts of the cluster, away from the more destructive interactions of the giant ellipticals.

A 2065. The Corona Borealis Cluster is an extremely challenging, albeit interesting cluster. With a distance of nearly 1.5 billion ly, it is one of the most remote objects in this catalogue. Of all the other clusters – only the A 1377 or the Ursa Major Cluster comes close at a billion light years. Since the brightest members are around 16.5 mag, it is effectively beyond the reach of most telescopes. However, with the remarkable sensitivity of CCD cameras, this cluster has become a reasonable target for more modest instruments.

A 2151. This is a rich, irregular cluster that bears many similarities to the Virgo Cluster, though it is almost 10 times more remote. It lacks a well-defined core region and many of the galaxies are late-type spirals. It contains several peculiar (IC 1182) and interacting pairs (IC 1179/81, NGC 6050 + IC 1179), plus dozens of other galaxies within range of a larger sized telescope.

A 2197 and *A 2199.* Both clusters are condensations near one end of the "Great Wall," and are located over 500 million ly away. Each cluster has a large cD galaxy in the core – NGC 6160 and NGC 6166 (Fig. 1.22), respectively [191].

A 3526. This is the Centaurus Cluster, a large, rich cluster of hundreds of galaxies. This impressive cluster is located approximately 170 million ly distant deep in the constellation of Centaurus. It is part of the Hydra-Centaurus Supercluster, which includes A 1060 and A 3574 [192,193]. In the dense core region lies NGC 4696, a 10.4 mag peculiar giant elliptical that is surrounded by a host of smaller systems. An interesting substructure is the Centaurus Chain, consisting of at least a dozen galaxies strewn in linear fashion in the NW section of the cluster. Much like the Markarian's Chain (Virgo Cluster), it contains an unusual assortment of galaxy types including interacting pairs (NGC 4622A/B) and the strange polar ring galaxy – NGC 4650A.

Chapter 10

Odd Stuff

Some targets or selections are "off the beaten path" [195], in the category "ultimate challenge" [196], or simply "exotic" [197]. The last chapter is a collection of observing programs or mere ideas, not fitting into one the schemes presented above – so more or less odd stuff. Feel free, to add your own creations!

Deep Sky Companions

Interesting views in the eyepiece are guaranteed in cases of chance alignments of a galaxy with a Milky Way object, such as a bright star, star cluster or planetary nebula [198]. In this category we will meet pairs ("cosmic companions") with a small angular, but extremely large linear separation.

Galaxies Near Bright Stars, Superimposed Stars

Galaxies near bright stars are very easy to find ("one-step starhopping"), but often difficult to observe. You must try to keep out the bright star in the field of view, while using a sufficiently high magnification. Such observations are just for fun and maybe set to the end of the observing session, when maximum dark adaptation (which can be killed by the star) is not further needed. The most prominent example might be "Mirach's Ghost," NGC 404, near β And (Fig. 10.1). But it is only second best when regarding the degree of separation (Table 10.1). The nice trio near ν Eri is shown in Fig. 10.2.

A frequent feature is a star superimposed on a galaxy. This often looks nice, but can cause some trouble too – it could be confused with a supernova! Suspicious candidates are M 108 (12 mag star near the center) and NGC 6207 (13.5 mag star near center), which is also the celebrated galaxy near M 13 (see Table 10.2).

Galaxies Near Nonstellar Galactic Objects

What about galaxies that apparently "pair up" with a well-known galactic Messier or NGC object? A classic example is M 13 and NGC 6207 – but also note that there are numerous fainter galaxies lying in the vicinity [199,200] (Table 10.2). If both "components" fit into the same (low power) field, the view can be striking. A far more challenging neighbor is IC 1296, a faint barred spiral near the planetary nebula M 57 in Lyra (Fig. 10.3). Numerous faint galaxies are visible close to the bright globular M 92 [201]. It can also be

Fig. 10.1. The peculiar S0 galaxy NGC 404 near Mirach (β And)

fascinating diversion to search for galaxies that lie "inside" popular open clusters. The Praesepe or "Bee hive" (M 44) has a nice collection [202], while the Pleiades (M 45) has only one, UGC 2838, which is a very difficult object. Numerous bright galaxies can be found in and around the Coma Berenices open cluster (Mel 111). Visual descriptions of galaxies near to bright stars or deep sky objects are presented in Table 10.3.

Another interesting variant are objects that have radically different distances. Often there are faint, distant galaxies in close proximity to much brighter (closer) ones. We have already mentioned the background galaxies near NGC 7331. Another interesting area is that around M 51 [203]. Try hunting for Keeler's six IC galaxies, which are located very close to the Whirlpool. The most difficult is the edge-on IC 4277 located only 8′ NW. The others are IC 4263, 4278, 4282, 4284, and 4285; all which all require 18″ or more aperture for observation.

Famous Names

Galaxies with proper names are obviously famous. Many like the "Tadpole" (Fig. 8.14) or the "Cartwheel," has been imaged with the Hubble Space Telescope, especially in the "Hubble Heritage Project." But forget seeing Hubble style "pretty pictures" when visual observing galaxies like those presented in Table 10.4 (descriptions in Table 10.6).

Fig. 10.2. The galaxy trio NGC 1618/22/25 in Eridanus, pretty near the bright starv Eri

Objects bearing famous names like Centaurus A (Fig. 7.8) or Cygnus A [204] make an interesting project. These are early radio source designations (brightest in a constellation). These "A-type" objects include a number of interesting galaxies (Table 10.5), and most can be observed visually, whereas Her A is challenging (see visual descriptions in Table 10.6).

Table 10.1. A sample of galaxies within 25′ of stars brighter than 4.5 mag (*d* = separation from star in arcmin)

Objects	*d*	PA	*V*	Type	Remarks
31 Leo			4.4	K4	Multiple star
NGC 3130	4.3	115	13.4	S0-a	
β Ari			2.6	A5	Sheratan
NGC 722	4.7	160	13.3	Sb	
β And			2.1	M0	Mirach
NGC 404	6.6	330	10.3	S0pec	Mirach's Ghost
λ Hya			3.6	K0	
NGC 3145	7.8	230	11.7	SBb-c	
η Peg			2.9	G2	Matar
NGC 7357	8.3	250	14.0	Sb	
ν Eri			3.9	B2	
NGC 1618	13.2	345	12.7	SBb	
NGC 1622	10.9	23	12.5	SBa-b	
NGC 1625	12.5	75	12.3	SBb	
β UMa			2.4	A1	Merak
NGC 3499	14.8	125	13.6	S0-a	

Table 10.2. Galaxies near nonstellar Milky Way objects (*d* = distance from center in arcmin)

Object	Type	Con	*d*	Galaxy	R.A.	Decl	*V*	*V'*	*a*×*b*	PA	Type
M 13	GC	Her	28	NGC 6207	16 43 03.7	+36 49 55	11.4	12.6	3.0×1.2	15	Sc
			15	IC 4617	16 42 08.1	+36 41 03	15.2	14.3	1.2×0.4	32	Sb
M 44	OC	Cnc	27	NGC 2624	08 38 09.6	+19 43 34	13.9	12.4	0.6×0.5	15	S
			20	NGC 2625	08 38 23.1	+19 42 58	14.3	13.4	0.7×0.6	45	E
			40	NGC 2647	08 42 43.0	+19 39 04	14.1	13.0	0.8×0.5	18	C
			5	IC 2388	08 39 56.5	+19 38 41	14.7	11.8	0.4×0.2	160	S
			29	IC 2390	08 41 51.7	+19 42 11	14.6	12.9	0.6×0.4	30	S
M 57	PN	Lyr	4	IC 1296	18 53 18.8	+33 03 59	14.3	14.0	1.1×0.8	80	SBbc
M 97	PN	UMa	48	M 108	11 11 29.4	+55 40 22	9.9	13.0	8.6×2.4	80	Sc
NGC 246	PN	Cet	26	NGC 255	00 47 47.3	–11 28 06	11.9	14.1	3.1×2.7	15	SBbc
NGC 288	GC	Scl	78	NGC 253	00 47 33.1	–25 17 15	7.3	12.9	29.0×6.8	52	SBc
NGC 6939	OC	Cep	40	NGC 6946	20 34 52.1	+60 09 12	9.0	14.0	11.5×9.8	57	SBc

Table 10.3. Visual descriptions of galaxies near to stars or deep sky objects (sorted by designation). Some galaxies were already described above: M 108 (Table 7.2), NGC 253, NGC 6946 (Table 7.5), NGC 1618/22/25 (Table 9.5)

Object	Description
NGC 255	Faint, small, round. Located 15′ NNE of NGC 246 (8″) Moderately bright, fairly large, elongated 4:3 NNW–SSE, 2.0′×1.6′, broad mild concentration. A 14 mag star lies 2.5′ ESE (SG 17.5″)
NGC 404	Bright, stellar nucleus with round, diffuse halo. Dark feature not visible. Bright star Mirach near (14″)
NGC 722	Very faint, very small, oval 3:2 NW–SE. Remarkable location as situated 7′ SE of β Ari in the same 220× field (SG 17.5″)
NGC 2624	Faint, very small, round, bright core. Forms a pair with NGC 2625 3.3′ ESE (SG 17.5″)
NGC 2625	Faint, extremely small, round. Appears similar to NGC 2624 3.3′ WNW but slightly smaller and fainter. Located at the W edge of M 44 (SG 17.5″)
NGC 2647	Faint, very small, round, 20 in. diameter, even surface brightness. Located 0.9′ NE of a 13 mag star at the east edge of M 44! (SG 17.5″)
NGC 3130	Fairly faint, small, round, weak concentration. The visibility of this galaxy is hindered by 31 Leo just 4.7′ WNW (SG 13″)
NGC 3145	Fairly faint, fairly small, nearly round, weak concentration. Overpowered by the glare of λ Hya 8′ NE (SG 13″)
NGC 3499	Fairly faint, very small, round, bright core, stellar nucleus. Located 14.8′ SE of β UMa (SG 17.5″)
NGC 6207	Bright, moderately large, elongated SSW–NNE (SG 13″) Fairly bright, very elongated 3:1 SSW–NNE, bright stellar nucleus, possible asymmetric appearance. Located 28′ NE of M 13. The noted stellar nucleus may be a superimposed foreground star (SG 17.5″)
NGC 7357	Very faint, small, round, 20 in. diameter. A 14 mag star is just off the NW edge 25 in. from center. View severely hampered by η Peg located 8′ NE! (SG 17.5″)
IC 1296	Extremely faint, small, round, very low surface brightness. Situated just 4′ NW of M 57! Located along the N side of a small rhombus of 10.5/12/13.5/13.5 mag stars with sides of 1.5′ (SG 17.5″)
IC 2390	Faint, small, stellar core (20″)
IC 4617	Very small, faint, and noticeably elongated (RJ 20″)

Additional Notes

Carafe. This is an unusual Seyfert Type II galaxy with an extensive, off-center ring (Fig. 10.4). It is part of a group of galaxies, including NGC 1595 and 1598 [217].

Minkowski's Object. This is an irregular galaxy and/or extragalactic HII region undergoing a major starburst. This intense star forming activity is thought to be the triggered by a radio jet emanating from the giant elliptical galaxy NGC 541 in the galaxy cluster A 194 [218].

Table 10.4. A (pretty subjective) sample of galaxies with famous names or designations

Object	Con	R.A.	Decl	V	$a \times b$	Type	Remarks
New 1	Cet	01 05 04.9	−06 12 45	11.6	4.4×3.3	SAB(rs)d	MCG -1-3-85
Minkowski's Object	Cet	01 25 44.4	−01 22 42	16.5	1.8×1.8	Irregular	Arp 133
Carafe	Cae	04 28 00.0	−47 54 46	12.5	1.5×1.2	Polar ring?	ESO 202-23, in group with NGC 1595/98
Das Rheingold	Vol	06 43 06.0	−74 14 11	13.5	1.5×0.9	Ring	ESO 34-11, Graham A
Integral Sign Galaxy	Cam	07 11 19.2	+71 50 12	12.9	3.2×0.3	Sd	UGC 3697
Coddington Nebula	UMa	10 28 22.4	+68 25 00	10.2	13.2×5.4	SAB(s)m	IC 2574, M 81 group
Reinmuth 80	Vir	12 32 28.1	+00 23 25	12.0	4.0×2.6	SB(rs)dm	NGC 4517A
GR 8	Vir	12 58 40.4	+14 13 03	14.4	1.1×0.7	Im	UGC 8091, beyond Local Group
Fath 703	Lib	15 13 48.1	−15 27 51	12.0	3.6×2.9	SA(s)d	NGC 5892
Hoag's Object	Ser	15 17 14.4	+21 35 08	16.0	0.3×0.3	Ring	PGC 54559
McLeish's Object	Pav	20 09 28.1	−66 13 00	15.1	1.0×0.3	S pec	ESO 105-26
Southern Integral Sign	Ind	22 14 44.5	−69 21 57	14.5	1.4×0.3	SBc:	IC 5173

Fig. 10.3. The faint Sc galaxy IC 1296 near the famous "Ring Nebula" M 57 in Lyra

GR 8. A long time local group suspect, recent studies now indicate it has a distance of 2.2 Mpc which place it beyond the group's outer boundary [207].

Hoag's Object. A nearly perfect ring galaxy formed by a collision billions of years ago (Fig. 1.28). Measuring about 100,000 light years in diameter and 600 million ly distant, it has been suggested that this is a polar ring galaxy seen "pole on" [219].

McLeish's Object. An interacting system first discovered by David McLeish in 1946. It consists of a large edge-on spiral galaxy with a highly disturbed NW side and a small, interloper galaxy [220].

Cygnus A. One of the most powerful radio sources in the sky, it was discovered in the early radio surveys back in the 1940s. Eventually its radio emissions were traced back to a distant giant elliptical galaxy located nearly a billion ly distant [221]. Its huge radio emission lobes are quite symmetrical and span about 500,000 ly. In optical images, the galaxy appears to have large central dust lane that bisects the galaxy in half. The quasar-like active galactic nucleus or AGN is thought to be powered by an immense black hole located in the core (Fig. 2.12). Much closer to us, Centaurus A (NGC 5128) displays a similar radio lobe structure, and it is also cut by a massive dust lane.

Table 10.5. All extragalactic "A-type" radio sources (3C = third catalogue of radio sources detected at Cambridge; Dist = distance in Mpc)

Source	3C	Optical	Type	R.A.	Decl.	V	$a \times b$	Dist	Remarks
And A		M 31	Ordinary	00 42 44.3	+41 16 09	3.4	178 × 63	0.76	Andromeda Nebula
Cen A		NGC 5128	Active	13 25 27.6	–43 01 09	7.0	25.7 × 20.0	4.6	
Cet A	3C 71	M 77	Seyfert	02 42 40.7	–00 00 48	8.9	7.1 × 6.0	18.4	Nearest Seyfert galaxy
Com A	3C 277.3	PGC 43882	Active	12 54 11.7	+27 37 33	15.0	0.7 × 0.3	107	Not in Coma Cluster
Cyg A	3C 405	MCG 7-41-3	cD	19 59 28.3	+40 44 02	15.1	0.5 × 0.3	240	Quasar core
For A		NGC 1316	Active	03 22 41.7	–37 12 30	8.2	13.5 × 9.3	16.3	Fornax Cluster
Her A	3C 348	MCG 1-43-6	cD	16 51 08.1	+04 59 33	17.0	0.3 × 0.2	550	
Hya A	3C 218	MCG -2-24-7	cD	09 18 05.7	–12 05 44	13.5	0.7 × 0.7	210	
Per A	3C 48	NGC 1275	Seyfert	03 19 48.1	+41 30 42	11.9	2.2 × 1.7	100	Perseus Cluster
Pic A		ESO 252-18	Active	05 19 49.7	–45 46 45	15.8	0.5 × 0.3	140	
UMa A	3C 231	M 82	Active	09 55 52.2	+69 40 47	8.4	11.2 × 4.3	3.7	Nearest peculiar galaxy
Vir A	3C 274	M 87	Active	12 30 49.3	+12 23 28	8.6	8.3 × 6.6	18.4	Virgo Cluster

Fig. 10.4. The "Carafe" ESO 202-23, a peculiar galaxy in Caelum (at lower left, in group with NGC 1595/98 north following)

Table 10.6. Visual descriptions of famous galaxies (sorted by name). Some objects were already described: M 31, M 51, M 77, M 82, and M 87 (Table 7.2), NGC 1316, NGC 5128 (Table 7.5), ESO 1-1 = NGC 2573 (Table 7.10), NGC 2537 (Table 8.23)

Object	Description
New 1	Faint, pretty small, very little brighter middle, little elongated 1.2′ × 1′ in PA 0°, averted vision makes it grow larger (SC 13.1″)
Carafe	1.4′ × 1′ haze, elongated N–S, with an irregularly shaped bar which is brighter and broader to the S. It appears grainy at 450× and two stellar knots can be seen, one near the centre and the other close SSE. A close binary star is 1′ ENE. NGC 1598 is a 1.2′ × 0.8′ haze elongated NW–SE which rises to a stellar nucleus and shows suggestions of two spiral arms curling away NW and SE especially with averted vision. NGC 1595 is a 0.8 × 0.6′ haze elongated N–S which rises to a stellar nucleus. Best viewed at 350× (Michael Kerr 25″)
Das Rheingold	Appears as a faint 0.5′ × 0.4′ haze elongated N–S, 0.7′ SSE of a 13 mag star. No structure is visible. With averted vision PGC 19455 is visible and PGC 19458 is just visible. Best viewed at 170× (Michael Kerr 8″) Appears as a fairly bright 0.4′ diameter hazy core surrounded by a 1.5′ × 0.7′ low surface brightness halo, elongated N–S, which is best seen with averted vision. A slight brightening can be seen on the SW and NE

Odd Stuff

	rim of the halo, and a 13 mag star is off the NW rim. Three other galaxies are nearby: ESO 34-11B, PGC 19455, and PGC 19458, and all are visible with direct vision as small faint round hazes. Best viewed at 270× (Michael Kerr 25″)
Minkowski's Object	Located in the NW part of the halo of NGC 541. Only about 40″ from the galaxies core, it is the middle object of three tiny dwarf galaxies in the area. This object was very difficult to detect at 400×, visible only as a very weak diffuse spot with adverted vision under good dark sky conditions (RJ 24″).
Integral Sign Galaxy	Very faint, extremely thin ghostly streak oriented WSW–ENE, at least 2.5′ × 0.3′. The surface brightness is very low and there is no significant concentration toward the center. I found this object difficult at 100× but it showed up fairly well at 220×. Unfortunately, those interesting curved tips were not detected (SG 17.5″)
Southern Integral-Sign	Observed at 28° elevation. Only visible with averted vision as an extremely faint 1.3′ × 0.2′ streak, elongated E–W. Careful observation shows a slightly brighter patch in the centre and a larger patch on the W end (IC 5173B). A 16 mag star is superposed on the N edge between these two patches and a fainter superposed star is glimpsed occasionally near the E end. Another 16 mag star is 0.8′ NW. Best viewed at 350× (Michael Kerr 25″)
Coddington Nebula	Faint, very large, elongated 5:2 SW–NE, 7.0′ × 2.5′, low surface brightness, no concentration. Four faint stars are near the N side. There is a fairly bright nonstellar HII region which is clearly visible at the NE end as a high surface brightness knot (SG 17.5″)
Reinmuth 80	Very faint, large, small brighter core. Appears as a very diffuse hazy region elongated SSW–NNE with no distinct boundaries (SG 17.5″)
GR 8	14 mag star at position with a very small glow extending northward, ~1′ × 0.5′ at best (Tom Polakis 13″)
Fath 703	Very faint but fairly large, round, 2.5′ diameter. Very low but uneven surface brightness (weak irregular concentration) with no distinct borders (SG 17.5″)
Hoag's Object	Core visible, but very faint, round (20″)
McLeish's Object	Best viewed at 450× with δ Pav positioned out of the field. McLeish A is a faint 1′ × 0.1′ streak, very slightly brighter toward the centre. McLeish B is visible as a very faint small round haze 1′ E. Interesting field with the strong contrast from δ Pav (Michael Kerr 25″)
Com A	Very faint, small, round, diffuse (20″)
Cyg A	Very difficult, averted vision (14″) Very faint, very small. In crowded field (20″)
Her A	Extremely faint, very small, between two stars (Frank Richardsen 20″)
Hya A	Fairly bright, almost stellar, diffuse halo (20″)
Per A	Extremely faint, extremely small, round (SG 8″) Fairly bright, fairly small, oval ~E–W, small bright core (SG 17.5″)
Pic A	Faint 0.6′ diameter round haze, which rises sharply to a stellar nucleus that is slightly fainter than the 16 mag star 0.5′ N. A 15 mag star is 1′ ESE and two 13 mag stars are 2′ NE and SW. The 13 mag star SW forms a triangle with two 15 mag stars and the E star is superposed on a small faint elongated galaxy. Best viewed at 450× (Michael Kerr 25″)

Appendix

Abbreviations

Here general abbreviations, as used in the text, are explained. Excluded: galaxy classification (Table 1.2), catalog designations and names (Chapter 3), publications (see next Appendices), or directions (e.g., NW).

AGN	Active galactic nucleus
BCD	Blue compact dwarf (galaxy)
BSO	Blue stellar object
CCD	Charge coupled device
CDM	Cold dark matter
DSS	Digital Sky Survey
fst	Faintest star
GA	Great attractor
GC	Globular cluster
GRB	Gamma ray burst
HII	HII region
HDF	Hubble Deep Field
HST	Hubble Space Telescope
LBV	Luminous blue variable
LF	Luminosity function
LG	Local Group
LPR	Light pollution reduction (filter)
LSB	Low surface brightness (galaxy)
OC	Open cluster
PN	Planetary nebula
POSS	Palomar Observatory Sky Survey
QSO	Quasi stellar object (quasar)
SC	Star cloud
SCT	Schmidt-Cassegrain telescope
SDSS	Sloan Digital Sky Survey
SSC	Super star cluster
UHC	Ultra high contrast (filter)
ULIRG	Ultra luminous infrared galaxy
WR	Wolf-Rayet (galaxy)
ZOA	Zone of avoidance

General Literature

Books on Galaxies and Related Subjects

Binney, J., Merrifield, M., *Galactic Astronomy*, Princeton University Press, Princeton, NJ, 1998
Bok, B. J., Bok, P. F., *The Milky Way*, Harvard University Press, Cambridge, MA, 1981
Combes, F., Boisse, P., Mazure, A., Blanchard, A., *Galaxies and Cosmology*, Springer-Verlag, Heidelberg, 2002
Elmegreen, D. M., *Galaxies and Galactic Structure*, Prentice-Hall Inc., Upper Saddle River, NJ, 1998
Ferris, T., *Galaxies*, Stewart, Tabori & Chang Publ., New York, NY, 1982
Hodge, P. W., *Galaxies*, Harvard University Press, Cambridge, MA, 1986
Hubble, E. P., *Realm of the Nebulae*, Dover Publ., Mineola, NY, 1958
Jones, M., Lambourne, R. (eds), *An Introduction to Galaxies and Cosmology*, Cambridge University Press, Cambridge, 2004
Keel, C. W., *The Road to Galaxy Formation*, Springer-Verlag, Heidelberg, 2002
Longair, M., *Galaxy Formation*, Springer-Verlag, Heidelberg, 1998
Sandage, A. R., *The Hubble Atlas of Galaxies*, Carnegie Inst., Washington, DC, 1961
Sandage, A. R., Bedke, J., *The Carnegie Atlas of Galaxies*, Carnegie Inst., Washington, DC, 1994
Simons, S., *Galaxies*, Harper Trophy, New York, NY, 1991
Waller, W. H., Hodge, P. W., *Galaxies and the Cosmic Frontier*, Harvard University Press, Cambridge, MA, 2003

Deep Sky Observing Books

Burnham, R. Jr., *Burnham's Celestial Handbook*, Vol. 3, Dover Publ., Mineola, NY, 1978
Houston, W. S., *Deep Sky Wonders*, Sky Publ. Corp., Cambridge, 1999
Kepple, G. R., Sanner, G., *The Night Sky Observers Guide*, Vol. 2, Willmann-Bell, Richmond, VA, 1998
Luginbuhl, C., Skiff, B. A., *Observing Handbook and Catalogue of Deep Sky Objects*, 2nd edition, SIGS Books and Multimedia, New York, 1999
Malin, D., Frew, D. J., *Hartung's Astronomical Objects for Southern Telescopes*, Cambridge University Press, Cambridge, 1994
Webb Society Deep Sky Observer's Handbook, Vol. 4: Galaxies, Vol. 5: Clusters of Galaxies, Vol. 6: Anonymous Galaxies, Vol. 7: The Southern Sky, Enslow Publishers, Berkeley Heights, NJ, 1979–87

Sky Atlases and Catalogs

Cragin, M., Bonanno, E., *Uranometria 2000.0 – Deep Sky Field Guide*, 2nd edition, Willmann-Bell, Richmond, VA, 2001
Hirshfeld, A., Sinnott, R., *Sky Catalogue 2000.0*, Vol. 2, Sky Publ. Corp., Cambridge, 1985
Sinnott, R. W., Perryman. M., *Millennium Star Atlas*, Vol. 3, Sky Publ. Corp./ESA, Cambridge, 1997

Strong, R. A., Sinnott, R. W., *Sky Atlas 2000.0 Companion*, 2nd edition, Sky Publ. Corp., Cambridge, 2000
Tirion, W., *Sky Atlas 2000.0*, Sky Publ. Corp., Cambridge, 1987
Tirion, W., Rappaport, B., Remaklus, W., *Uranometria 2000.0 – Deep Sky Atlas*, 2 Vol., 2nd edition, Willmann-Bell, Richmond, VA, 2001

Magazines (with frequent articles on galaxies)

Astronomy, Kalmbach Publication (USA)
Astronomy & Geophysics, Royal Astronomical Society (UK)
Astronomy and Space, Astronomy and Space Magazine (AUS)
Astronomy Now, Pole Star Publications Ltd (UK)
Deep Sky Observer, Webb Society (UK)
Journal of the BAA, British Astronomical Association (UK)
Journal of the RASC, Royal Astronomical Society of Canada (CAN)
Popular Astronomy, The Society for Popular Astronomy (UK)
Sky & Telescope, Sky Publishing Corp. (USA)
Sky and Space Magazine, Sky and Space Publishing (AUS)

Digital Sources

Sky Mapping Software

Chartes du Ciel: www.stargazing.net/astropc
Guide, Project Pluto: www.projectpluto.com
MegaStar, E.L.B. Software: home.flash.net/~megastar
Redshift, Maris Technologies: www.redshift.maris.com
SkyChart, Southern Stars Software: www.southernstars.com
SkyMap Pro, SkyMap Software: www.skymap.com
StarryNight Pro, Space Holdings Inc.: www.starrynight.com
The Sky, *RealSky*, Software Bisque: www.bisque.com
Xephem, Clearsky Institute: www.clearskyinstitute.com/xephem

Internet Databases

Centre des Donneés Astronomique de Strasbourg (CDS), catalogues: cdsweb.u-strasbg.fr/cats/cats.html
Lyon-Meudon Extragalactic Database (LEDA): leda.univ-lyon1.fr
Mikkel Steine's *Deep Sky Browser*, collection of deep sky catalogues: messier45.com
NASA Extragalactic Database (NED): nedwww.ipac.caltech.edu
Revised New General Catalogue and Index Catalogue: www.klimaluft.de/steinicke/ngcic/rev2000/Explan.htm
Saguaro Astronomy Club (SAC), Deep Sky Database: www.saguaroastro.org
SIMBAD/Aladin: simbad.u-strasbg.fr/Simbad

Useful Links

Amastro mailing list: groups.yahoo.com/group/amastro
Anglo Australien Telescope (AAT), images: www.aao.gov.au/images
Astrophysics Data System (ADS): adsabs.harvard.edu/article_service.html
Digitized Sky Survey (DSS): archive.eso.org/dss/dss
Hubble Space Telescope (HST), images: www.seds.org/hst/hst.html
Internet Amateur Astronomers Catalog (IAAC), visual observations: www.visualdeepsky.org
National Space Science Data Center (NDSSC), photo gallery: nssdc.gsfc.nasa.gov/photo_gallery
NGC/IC Project: www.ngcic.org
Optical Master Catalog: heasarc.gsfc.nasa.gov/W3Browse/all/optical.html
Preprint-Server: www.stsci.edu/astroweb/cat-preprint.html
Skyview Advanced: skyview.gsfc.nasa.gov/cgi-bin/skvadvanced.pl
Students for the Exploration of Space (SEDS): www.seds.org
Webb Society: www.webbsociety.freeserve.co.uk
Website of the Author (data, articles): www.klima-luft.de/steinicke

References

The following abbreviations are used:

S&T	Sky & Telescope
DSO	Deep Sky Observer
DSM	Deep Sky Magazine
MNRAS	Monthly Notices of the Royal Astronomical Society
Sci. Am.	Scientific American
Astrophys. J.	Astrophysical Journal
Astrophys. J. Suppl.	Astrophysical Journal Supplement
Astron. J.	Astronomical Journal
Astron. Astrophys.	Astronomy & Astrophysics

[1] Lucentini, J., The Mysteries of Galaxy Spirals, S&T 9/2002, 29
[2] Voit, G. M., The Rise and Fall of Quasars, S&T 5/1999, 40
[3] Ford, H., Tsvetanov, Z. I., Massive Black Holes in the Hearts of Galaxies, S&T 6/1996, 28
[4] Deckel, A., Ostriker, J. P., Formation of Structure in the Universe, Cambridge University Press, Cambridge, 1999
[5] Hogan, C. J., The Little Book of the Big Bang, Springer-Verlag, Heidelberg, 1998
[6] Livio, M., The Accelerating Universe, John Wiley & Sons, New York, 2000
[7] King, I. R., Gilmore, G., van der Kruit, P. C., The Milky Way as a Galaxy, University Science Books, Mill Valley, CA, 1990
[8] Kraan-Korteweg, R. C., Lahav, O., Galaxies Behind the Milky Way, Sci. Am. 10/1998, 50
[9] Binney, J., The Evolution of Our Galaxy, S&T 3/1995, 20
[10] Trimble, V., Parker, S., Meet the Milky Way, S&T 1/1995, 26

[11] Roth, J., Primack, J. R., Cosmology: All Sewn Up or Coming Apart at the Seams? S&T 1/1996, 20

[12] Hodge, P., The Extragalactic Distance Scale: Agreement at Last? S&T 10/1993, 16

[13] Parker, B., Discovery of the Expanding Universe, S&T 9/1986, 227

[14] Davies, P., Everyone's Guide to Cosmology, S&T 3/1991, 250

[15] Odenwald, S., Fienberg, R. T., Galaxy Redshifts Reconsidered, S&T 2/1993, 31

[16] Freedman, W. L., Turner, M. S., Cosmology in the New Millenium, S&T 10/2003, 30

[17] MacRobert, A., Understanding Celestial Coordinates, S&T 12/1995, 38

[18] Skiff, B. A, Galaxies and their Magnitudes, Part I: DSM 31, 50 (1990); Part II: DSM 32, 50 (1990); Part III: DSM 33, 50 (1990); Part IV: DSM 34, 50 (1991), Part V: DSM 35, 50 (1991)

[19] Clark, R. N., What Magnitude Is It? S&T 1/1997, 118

[20] Shapley-Ames catalog: nedwww.ipac.caltech. edu/level5/Shapley_Ames/frames.html

[21] van den Bergh, S., Galaxy Morphology and Classification, Cambridge University Press, Cambridge, 1998

[22] Lake, G., Understanding the Hubble Sequence, S&T 5/1992, 515

[23] Bertola, F., What Shape are Elliptical Galaxies? S&T 5/1981, 380

[24] de Vaucouleurs, G. A., et al., Third Reference Catalogue of Bright Galaxies, Springer-Verlag, Heidelberg, 1991

[25] Beck, S. C., Dwarf Galaxies and Starbursts, Sci. Am. 6/2000, 66

[26] Bothun, G. D., Beyond the Hubble Sequence, S&T 5/2000, 36

[27] Block, D. L., The Duality of Spiral Structure, S&T 1/2001, 48

[28] 2MASS Galaxy Morphology: www.ipac.caltech.edu/2mass/gallery/galmorph/2mass_galmorp.html

[29] Steinicke, W., Superthin Galaxies – Objects Sharp Like a Razor Blade, DSO 125, 24 (2001)

[30] Finkbeiner, A., Active Galactic Nuclei: Sorting Out the Mess, S&T 8/1992, 138

[31] Schramm, D. N., The Origin of Cosmic Structure, S&T 8/1991, 140

[32] Harris, W. E., Globular Clusters in Distant Galaxies, S&T 2/1991, 148

[33] Djorgovski, S. G., The Dynamic Lifes of Globular Clusters, S&T 10/1998, 38

[34] Dalcanton, J., The Overlooked Galaxies, S&T 4/1998, 28

[35] Bothun, G., The Ghostliest Galaxies, Sci. Am. 2/1997, 56

[36] Dressler, A., Galaxies Far Away & Long Ago, S&T 4/1993, 22; see also: Observing Galaxies Through Time, S&T 8/1991, 126

[37] Parker, S., Roth, J., The Hubble Deep Field, S&T 5/1996, 48

[38] Bertin, G., Lin, C. C., Spiral Structures in Galaxies, MIT Press, Cambridge, MA, 1996

[39] Martin, P., Friedli, D., At the Hearts of Barred Galaxies, S&T 5/1999, 32

[40] Weil, T. A., Another Look at Cosmic Distances, S&T 8/2001, 62

[41] Sandage, A. R., The Light Travel Time and the Evolutionary Correction to Magnitudes of Distant Galaxies, Astrophys. J. 134, 916 (1961)

[42] Bechtold, J., Shadows of Creation: Quasar Absorption Lines and the Genesis of Galaxies, S&T 9/1997, 28

[43] Disney, M., A New Look on Quasars, Sci. Am. 10/1994, 28

[44] West, M., Galaxy Clusters: Urbanization of the Cosmos, S&T 1/1997, 30

[45] Lake, G., Cosmology of the Local Group, S&T 12/1992, 613

[46] Naeye, R., The Newest Closest Galaxy, S&T 2/2004, 22; see also: astro.u-strasbg.fr/images_ri/canm_e.html

[47] Jayawardhana, R., Our Galaxy's Nearest Neighbour, S&T 5/1998, 42

[48] Humason, M. L., Mayall, N. U., Sandage, A. R., Redshifts and Magnitudes of Extragalactic Nebulae, Astrophys. J. 61, 97 (1956)

[49] Chincarini, G., Rood, H. J., The Cosmic Tapestry, S&T 5/1980, 364

[50] Geller, M., Mapping the Universe, S&T 8/1991, 134

[51] Schilling, G., Cosmology's Treasure Map: The 2dF Galaxy Redshift Survey, S&T 2/2003, 32

[52] Knapp, G. R., Mining the Heavens – The Sloan Digital Sky Survey, S&T 8/1997, 40

[53] Roth, J., When Galaxics Collide, S&T 3/1998, 48

[54] Arp, H., Seeing Red: Redshifts, Cosmology and Academic Science, Apieron 1998 (review: Barry F. Madore, S&T 6/1999, 87)

[55] Petersen, C. C., The Universe through Gravity's Lens, S&T 9/2001, 32

[56] Schild, R. E., Gravity Is My Telescope, S&T 4/1991, 375

[57] Morgan, W. W., et al., cD Galaxies in Poor Clusters, Astrophys. J. 199, 545 (1975); Astrophys. J. 211, 309 (1977)

[58] Abell, G. O., The Distribution of Rich Clusters of Galaxies, Astrophys. J. Suppl. 3, 211 (1958)

[59] Abell, G. O., Clusters of Galaxies, in: Galaxies and the Universe, A. Sandage (ed.), Chicago University Press, Chicago, IL, 1975

[60] Henry, P. J., Briel, U. G., Boehringer, H., The Evolution of Galaxy Clusters, Sci. Am. 12/1998, 52

[61] Bahcall, N. A., Soneira, R. M., A Supercluster Catalog, Astrophys. J. 277, 27 (1984)

[62] Zucca, E., et al., All-sky Catalogs of Superclusters of Abell-ACO-Clusters, Astrophys. J. 407, 470 (1993)

[63] Glyn Jones, K., Messier's Nebulae and Star Clusters, 2nd edition, Cambridge University Press, Cambridge, 1991

[64] Mallas, J. H., Kreimer, E., The Messier Album, Sky Publ. Corp., Cambridge, 1978

[65] Frommert, H., Messier catalogue: www.seds.org/messier

[66] O'Meara, S. J., Deep Sky Companions: The Messier Objects, Cambridge University Press, Cambridge, 2000

[67] Moore, P., Beyond Messier: The Caldwell Catalog, S&T 12/1995, 38

[68] O'Meara, S. J., Deep Sky Companions: The Caldwell Objects, Sky Publ. Corp., Cambridge, 2002

[69] Dreyer, J. L. E., New General Catalogue, Index Catalogue, Second Index Catalogue, Royal Astronomical Society, London, 1962

[70] Brierley, P., Observing the Herschel 400, DSO 123, 1 (2001)

[71] Steinicke, W., Digital Deep Sky Data, Visual Observing and the NGC/IC Project: www.klima-luft.de/steinicke/Deep-Sky/deep-sky_e.htm

[72] Sulentic, J. W., Tifft, W. G., The Revised New General Catalogue of Nonstellar Astronomical Objects, University of Arizona Press, Tucson, 1977

[73] Sinnott, R. W., NGC2000.0, Sky Publ. Corp., Cambridge, 1988

[74] Steinicke, W., Revised New General Catalogue and Index Catalogue: www.ngcic.org, or for the latest update: www. klima-luft/steinicke/ngcic/rev2000/Explan.htm

[75] Gottlieb, S., Restoring Order to the Deep Sky, S&T 11/2003, 113

[76] Humphreys, R. M., Thurmes, P. M., A Billion Stars, A Few Million Galaxies, S&T 5/1994, 32

[77] Nilson, P., Uppsala General Catalogue of Galaxies, Uppsala Astron. Obs. 1973

[78] Higgins, D., The Uppsala General Catalogue and its Place in History, DSM 28, 18 (1989)

[79] Bunge, R., The Uppsala General Catalogue: An Overdue Review, DSM 30, 30 (1990)

[80] Nilson, P., Catalogue of Selected Non-UGC Galaxies, Uppsala Astron. Obs. 1974

[81] Sandage, A. R., Tammann, G. A., A Revised Shapley–Ames Catalog of Bright Galaxies, Carnegie Inst., Washington, DC, 1987

[82] Lauberts, A., The ESO/Uppsala Survey of the ESO(B) Atlas, ESO/Garching, 1982

[83] Lauberts, A., Vilenkin, E. A., The Surface Photometry Catalogue of ESO/Uppsala Galaxies, ESO/Garching, 1989

[84] Corwin, H. C., de Vaucouleurs, A., de Vaucouleurs, G., Southern Galaxy Catalogue, University of Texas Press, Austin, TX, 1985

[85] South-Equatorial Galaxy Catalogue: spider.ipac.caltech.edu/staff/hgcjr

[86] Dixon, R. S., Sonneborn, G., A Master List of Nonstellar Optical Astronomical Objects, Ohio State University Press, Columbus, OH, 1980

[87] Skiff, B. A., Delving into Deep Sky Data, S&T 3/2002, 50

[88] Databases: www.stsci.edu/astroweb/yp_center.html.

[89] van den Bergh, S., Luminosity Classification of Dwarf Galaxies, Astron. J. 71, 922 (1966)

[90] Tully, R. B., Fisher, J. R., Catalog of Nearby Galaxies, Cambridge University Press, Cambridge, 1988

[91] Arp, H. C., Madore, B. F., A Catalogue of Southern Peculiar Galaxies and Associations, Vol. 2, Cambridge University Press, Cambridge, 1987

[92] Veron-Cetty, M. P., Veron, P., Catalog of quasars, BL Lacertae objects and AGN: vizier.u-strasbg.fr/viz-bin/VizieR?-source=VII/235

[93] Burbidge, G., Hewitt, A., A Catalog of Quasars Near and Far, S&T 12/1994, 32

[94] Hickson. P., Atlas of Compact Groups of Galaxies, Gordon and Breach, Newark, NJ, 1994

[95] Skiff, B. A., The Curious Case of Copeland's Septet, DSM 2, 30 (1983)

[96] Steinicke, W., Extragalactic Objects Discovered as Variable Stars, Webb Society Publ., 2002

[97] Privett, G., Parsons, P., The Deep Sky Observer's Year, Springer-Verlag, Heidelberg, 2001

[98] Inglis, M., Field Guide to the Deep Sky Objects, Springer-Verlag, Heidelberg, 2001

[99] Coe, S., Deep Sky Observing: The Astronomical Tourist, Springer-Verlag, Heidelberg, 2000

[100] Eicher, D., The Universe from your Backyard, Kalmbach Publ., Waukesha, WI, 1992; see also: Stars & Galaxies, Kalmbach Publ., Waukesha, WI, 1997; and: Deep Sky Observing with Small Telescopes, Kalmbach Publ., Waukesha, WI, 1992

[101] Ashford, A. R., Star Catalogs for the 21st Century, S&T 7/2001, 65
[102] Dickinson, T., Dyer, A., The Backyard Astronomer's Guide, 2nd ed., Sky Publ. Corp., Cambridge, 2002
[103] Buta, R., The Dream Period, Part V: The Spiral Menagerie, DSO 131, 1 (2003)
[104] MacRobert, A., Beating the Seeing, S&T 4/1995, 40
[105] MacRobert, A., Pepin, M. B., Your Basic Eyepiece Set, S&T 4/1996, 38
[106] Sinnott, R. W., Rx for the Astigmatic Skywatcher, S&T 9/1995, 46
[107] Thessin, R., Crawford, D. L., Your Home Lighting Guide, S&T, 4/2002, 40
[108] Harrington, P., Nebula Filters for Light-Polluted Skies, S&T 7/1995, 38
[109] Dickinson, T., The Barlow Lens: More Power to You, S&T 67/1997, 59
[110] Regen, D. N., Unit-Power Finders, S&T 6/1996, 48
[111] Clark, R., N., Visual Astronomy of the Deep Sky, Sky Publ. Corp., Cambridge, 1990
[112] Clark, R., N., A Visual Tour to M 101, S&T 6/1993, 102
[113] Clark, R., N., How Faint You Can See? S&T 4/1994, 106
[114] Schaeffer, B. E., How Faint Can You See? S&T 3/1989, 332
[115] ODM (Mel Bartels): www.efn.org/~mbartels/aa/visual.html
[116] Ferris, W. D., Dark Skies Rule, S&T 8/2003, 62
[117] Schaeffer, B. E., Your Telescope Limiting Magnitude, S&T 11/1989, 522
[118] Bortle, J. E., Introducing the Bortle Dark-Sky Scale, S&T 2/2001, 126
[119] Mullaney, J., 111 Deep Sky Wonders for Light-Polluted Skies, S&T 4/2003, 110
[120] Harrington, P. S., The Deep Sky: An Introduction, Sky Publ. Corp., Cambridge, 1998
[121] Garfinkle, R., Star-Hopping – Your Visa to Viewing the Universe, Cambridge University Press, Cambridge, 1994
[122] McRobert, A., Star-Hopping for Backyard Astronomers, Sky Publ. Corp., Cambridge, 1993
[123] Jackiel, R., Tips and Techniques for Sketching the Deep Sky, S&T 3/2003, 115
[124] Freeman, J. R., Messier Surveys by the Dozen, S&T 3/2001, 108
[125] Freeman, J. R., Refractor Red Meets the Herschel 400, S&T 5/1999, 114
[126] Kay, J., A Visual Atlas of the Magellanic Clouds, Webb Society Publ., 2001
[127] MacRobert, A., A Galaxy-Hop in Leo, S&T 4/1997, 56
[128] Skiff, B. A., Pegasus – Explore the Square's Realm of the Unknown Galaxies, DSM 32, 12 (1990)
[129] Sandström, J., Galaxy Rambling – Canes Venatici et Environs, DSO 124, 17 (2001)
[130] Radloff, M., The Galaxies of Canes Venatici, DSM 22, 8 (1988)
[131] Skiff, B. A., Galaxies in Triangulum, DSM 4, 18 (1983)
[132] Steinicke, W., Galaxies in Leo Minor – "The Amateur Deep-Field": www.klima-luft.de/steinicke/Projekte/LMi/LMi_e.htm
[133] Ostuno, E., A Survey of Equuleus, DSM 32, 30 (1990)
[134] Ostuno, E., Galaxies Near the Winter Milky Way, DSM 29, 8 (1989); Galaxies along the Summer Milky Way, DSM 31, 18 (1990)
[135] McNeil, J., The Other Cygnus – Faint Galaxies of the Swan, S&T 7/2000, 117
[136] Gottlieb, S., Galaxies in Orion, DSM 36, 31 (1991)

[137] Hewitt-White, K., Galaxy Groups near Markab, S&T 11/2003, 119
[138] Skiff, B. A., The Other Lyra, DSM 27, 8 (1989)
[139] Kemble, L. J., Fifty-to-the-pole: The Observing Project of the Lifetime, DSM 5, 20 (1987)
[140] Van den Bergh, S., The Galaxies of the Local Group, J. R. Astron. Soc. Canada, 62, 145 (1968)
[141] Freemann, J. R., Observing Faint Nearby Galaxies, S&T 10/2002, 103
[142] Polakis, T., Observing the Local Group, DSM 36, 12 (1991)
[143] Goldstein, A., Observing the Local Group, DSM 4, 12 (1983)
[144] Ling, A., Dwarf Observations for "Dwarf" Telesopes, DSM 34, 46 (1991)
[145] Buta, R., The Dream Period, Part III: The IC 342/Maffei Group, DSO, 120, 1 (2000)
[146] Jackiel, R., Behind the Veil: Exploring the IC 342 group, S&T 3/2000, 126
[147] Polakis, T., Observing the M 81 Group of Galaxies, DSM 37, 16 (1991)
[148] Holmberg low surface brightness galaxies: c3po.cochise.cc.az.us/astro/deepsky04.htm
[149] Meylan, G., Brandl, B., 30 Doradus – Birth of a Star Cluster, S&T 3/1998, 40
[150] Mitchell, L., The M 31 Challenge, S&T 11/1997, 106
[151] Hodge, P. W., Atlas of the Andromeda Galaxy: nedwww.ipac.caltech.edu/level5/AndROMEDA_Atlas/frames.html
[152] Skiff, B. A., All About M 31, DSM 8, 8 (1984)
[153] Higgins, D., The M 31 Globular System, DSM 32, 24 (1990)
[154] Jackiel, R., A Tour of Extragalactic Globulars, S&T 10/2001, 115
[155] Jakiel, R., Globular Clusters in M 33: www.angelfire.com/id/jsredshift/gcm33.htm
[156] Chandar, R., Bianchi, L., Ford, H. C., Star Clusters in M 33. Part IV: A New Survey From Deep HST Images, Astron. Astrophys. 366, 498 (2001)
[157] Harris, W. E., Globular Cluster Systems in Galaxies beyond the Local Group: ned.ipac.caltech. edu/level5/Harris/Harris_contents.html
[158] Thompson, G. D., Bryan, J., Supernovae Search Charts and Handbook, Cambridge University Press, Cambridge, 1990; latest supernovae: www.rochesterastronomy.com/supernova.html
[159] Hanson, T., The "Deepest" Deep Sky Objects, DSM 34, 32 (1991)
[160] Craine, E. R., A Handbook of Quasistellar and BL Lacertae Objects, Pachart Publ. House, Tucson, 1977
[161] Steinicke, W., Catalog of Bright Quasars and BL Lacertae Objects: www.klima-luft.de/steinicke/KHQ/KHQ_e.htm
[162] Polakis, T., A Sampling of Edge-on Galaxies, S&T 11/2001, 122; see also: www.psiaz.com/polakis/edgeons/edgeon.html
[163] Skiff, B. A., Exploring the Hubble Sequence by Eye, S&T 5/2000, 120
[164] Goldstein, A., Observing the Morphology of Galaxies, DSM 28, 19 (1989)
[165] Steinicke, W., Catalog of Galaxy Groups: www.klima-luft.de/steinicke/KDG/KDG_e.htm
[166] Paul, M., Atlas of Galaxy Trios, DSO 122 (2000); southern extension: DSO 128, 10 (2002)
[167] Kay, J., A Night at the Bedford 30-inch, DSO 126 (2001)

[168] Goldstein, A., Observing Interacting Galaxies, DSM 18, 8 (1987)

[169] Jackiel, R., Galaxy Bridges, Tails, and Rings, S&T 11/2000, 122; More Bridges, Tails and Rings, S&T 5/2001, 124

[170] Gottlieb, S., Quintets, Sextets, and Septets: Exploring Hickson Compact Groups, S&T 3/1999, 110

[171] Morales, R. J., Visual Observations of Hickson's Galaxy Groups, DSO 123, 18 (2001)

[172] Freeman, J. R., Exploring Stephan's Quintet, S&T 9/2000, 117

[173] Corder, J., Gottlieb, S., NGC 7331 and Its Ambiguous Galaxies, DSM 16, 14 (1986)

[174] Hewitt-White, K., Two Galaxy Fields in Pegasus, S&T 12/2002, 114

[175] Hewitt-White, K., Great Galaxies, By Gemini! S&T 3/2001, 113

[176] Lamperti, A., Gottlieb, S., Observing Galaxy Chains, DSO 130, 12 (2002)

[177] Corder, J., The IC 698 Galaxy Group, DSM 17, 22 (1986)

[178] Morales, R. J., The NGC 383 Group, DSO 124, 20 (2001)

[179] Hewitt-White, K., A Circlet of Galaxies in Pisces, S&T 12/2003, 119

[180] Bunge, R., Tracking Down the NGC 4005 Group, DSM 26, 18 (1989)

[181] Meketa, J., Observing the Pegasus I Cluster, DSM 4, 21 (1983)

[182] Hewitt-White, K., The Pegasus I Galaxy Cluster, S&T 11/2001, 125

[183] Hewitt-White, K., Triangular Targets: Exploring Abell 262, S&T 1/2003, 115

[184] Gottlieb, S., On the Edge – the Corona Borealis Galaxy Cluster, S&T 5/2000, 128

[185] Gottlieb, S., The Pisces-Perseus Supercluster, S&T 1/2002, 133

[186] Buta, R., The Dream Period, Part IV: A Grand Tour of the Hercules Supercluster, DSO 128, 10 (2002)

[187] Skiff, B. A., The Coma Berenices and Abell 1367 Galaxy Clusters, DSM 10, 6 (1985)

[188] Morales, R. J., Abell 119, DSO 112, 21 (1998)

[189] Hewitt-White, K., The Ghosts of 891 Andromeda Way, S&T 12/2000, 125

[190] Morales, R. J., Abell 1314, DSO 130, 7 (2002)

[191] Hewitt-White, K., Two Galaxy Clusters in Hercules, S&T 6/2000, 115

[192] Corder, J., Observing the Centaurus Galaxy Cluster, DSM 22, 22 (1988)

[193] Gottlieb, S., The Hydra-Centaurus Supercluster, S&T 4/2002, 101

[194] Gottlieb, S., Exploring a Southern Galaxy Cluster, S&T 10/1999, 123

[195] Gottlieb, S., Off the Beaten Path: www.angelfire.com/id/jsredshift/offpath.htm

[196] Jakiel, R.,The Ultimate Deep Sky Challenge? DSO 112, 24 (1998)

[197] Eicher, D. J., Twenty-One Exotic Deep Sky Objects, DSM 9, 6 (1984)

[198] Gottlieb, S., Celestial Odd Couples, S&T 7/2003, 117

[199] Bunge, R., A Night of Galaxies Near M 13, DSM 27, 16 (1989)

[200] Hewitt-White, K., Snaring Galaxies Near M 13, S&T 6/2001, 117

[201] Hewitt-White, K., A Chain of Galaxies near M 92, S&T 7/2001, 114

[202] Hewitt-White, K., A Hive of Galaxies in Cancer, S&T 3/2003, 120

[203] Bunge, R., Exploring the Region of M 51, DSM 30, 16 (1990)

[204] Steinicke, W., Cygnus A – Observing an Exceptional Radio Galaxy: www.klima-luft.de/steinicke/Artikel/cyga/cyga_e.htm

[205] Jakiel, R., Observations of the Dwarf Seyfert Galaxy NGC 4395, DSO 111, 4 (1998)

[206] Rowan-Robinson, M., The Cosmological Distance Ladder, Freeman & Co., New York, 1985

[207] van den Bergh, S., The Galaxies of the Local Group, Cambridge University Press, Cambridge, 2000

[208] Florido, E., et al., Corrugations in the discs of spiral galaxies – NGC 4244 and 5023, MNRAS 251, 193 (1991)

[209] Sparke, L., Gallagher, J., Galaxies in the Universe, Cambridge University Press, Cambridge, 2000

[210] van den Bergh, S., Galaxy Morphology and Classification, Cambridge University Press, Cambridge, 1998

[211] Carter, D., et al., Minor axis rotation and the intrinsic shape of the shell galaxy NGC 3923, MRNAS 294, 182 (1998)

[212] Bridges, T., et al., B–R colours of the globular clusters in NGC 6166 (A 2199), MNRAS 281, 1290 (1996)

[213] Ikebe, Y., et al., RXTE observation of NGC 6240: search for an obscured active nucleus, MNRAS 316, 433 (2000)

[214] White, R., et al., Seeing galaxies through thick and thin: Optical opacity measures in overlapping galaxies, Astron. J. 542, 761 (2000)

[215] Webb, S., Measuring The Universe: The Cosmological Distance Ladder, Springer-Verlag, Heidelberg, 2001

[216] Souza, A., Quintana, H., The NGC 4782/3 dumbbell massive group, Astron. J. 99, 1065 (1990)

[217] Hawarden, T. G., et al., The Carafe: a peculiar ringed Seyfert galaxy and its companions, MNRAS 186, 495 (1979)

[218] van Breugel, W., et al., Minkowski's Object: A starburst triggered by a radio jet, Astron. J. 293, 83 (1985)

[219] Horellou, C., et al., The CO emission of ring galaxies, Astron. Astrophys. 298, 743 (1995)

[220] Diaz, R., et al., Study of McLeish's Interacting Object, Astron. J. 119, 111 (2000)

[221] Burke, B., Graham-Smith, F., An Introduction to Radio Astronomy, Cambridge University Press, Cambridge, 1997

[222] Alonso-Herrero, A., et al., Nuclear star formation in the hotspot galaxy NGC 2903, MNRAS 322, 757 (2001)

[223] Schilling, G., Catching Gamma-Ray Bursts on the Wing, S&T 3/2004, 33

[224] French, S., Markarian's Chain, S&T 5/2004, 88

[225] Gottlieb, S., HII Regions Galore in M 101, S&T 6/2004, 89

[226] Cowen, R., A Stab in the Dark, S&T 7/2004, 43

[227] Waller, W. H., Redesigning the Milky Way, S&T 9/2004, 51

[228] Mobberley, M. P., A Trio of Supernova Hunters, S&T 10/2004, 111

[229] Charlton, J. C., et al., Galaxy Train Wrecks, S&T 11/2004, 31

[230] Whitman, A., Digging Deep in M 33, S&T 12/2004, 93

[231] Strauss, M. A., Knapp, G. R., The Sloan Digital Sky Survey, S&T 2/2005, 34

[232] O'Meara, S. J., M 102: Mystery Solved, S&T 3/2005, 78

[233] Miller, M. C., et al., Supermassive Black Holes: Shaping Their Surroundings, S&T 4/2005, 43
[234] Naeye, R., Meet the Milky Way's Newfound Neighbours, S&T 7/2005, 16
[235] Levy, D. H., Seeing Einstein's Gravity Lens, S&T 7/2005, 108
[236] Roth, J., Spectacles for Spectacular Skies, S&T 9/2005, 31
[237] Wanjek, C., Lasik Eye Surgery for the Amateur Astronomer, S&T 9/2005, 36
[238] Aguirre, E. L., German Amateur's Supernova in M 51, S&T 11/2005, 103
[239] Madore, B. F., De Paz, A. G., Discovery among the Disks, S&T 11/2005, 41
[240] Aguirre, E. L., Amateur Team Finds 100 Supernovae, S&T 11/2005, 101
[241] Roth, J., Setback for Alternative Cosmology, S&T 12/2005, 22
[242] French, S., Island Universe, S&T 12/2005, 77
[243] Devitt, T., Galactic survey reveals a new look for the Milky Way: www.news.wisc.edu/11405.html
[244] Lamperti, A., Extragalactic Globular Clusters, DSO 138, 23 (2005)
[245] Morales, R., NGC 128 and its Companions, DSO 137, 21 (2005)
[246] Brazell, O., The Gravitational Lens – A Visual & CCD Challenge, DSO 135, 27 (2004)
[247] Lamperti, A., Observing Some Galaxy Trios, DSO 134, 3 (2004)
[248] Sandström, J., Visual Observations of the Galaxy Cluster Pegasus I, DSO 134, 12 (2004)
[249] Morales, R. J., Abell 194 – A Guided Tour, DSO 133, 1 (2004)
[250] Summerfield, D., The Major Galaxies of Leo Minor, DSO 133, 12 (2004)
[251] Sandström, J., Observations of the Galaxy Cluster AGC 1367 & the Galaxy group NGC 4005 in Leo, DSO 133, 15 (2004)
[252] Zepf, S.E., Ashman, K.M., The Unexpected Youth of Globular Clusters, Sci. Am. 289, 28 (2003)
[253] Wakker, B. P., Richter P., Our Growing, Breathing Galaxy, Sci. Am. 290, 38 (2004)
[254] Lineweaver, C. H., Davis, T. M., Misconceptions About the Big Bang, Sci. Am. 3/2005, 36
[255] Longo, G., Capaccioli, M., Busarello, G. (eds), Morphological and Physical Classification of Galaxies, Kluwer Academic Publishers, Dordrecht, 1992
[256] Block, D. L., et al. (eds), Toward a New Millenium in Galaxy Morphology, Kluwer Academic Publishers, Dordrecht, 2000
[257] de Veauclouleurs, G., Tilt Criteria and Direction of Rotation of Spiral Galaxies, Astrophys. J. 127, 492 (1958)

List of Tables

Section I

Section II

Section III

Figured Objects

Object	Fig.
A 151	9.7
A 426	2.11
A 1656	1.2
A 1689	2.14
A 2029	2.13
A 2151	2.10
Arp 220	1.37
AWM 7	9.6
Coal Sack	1.5
Copeland's Septett	3.9
Cygnus A	2.12
Double Quasar	8.10
Draco Dwarf	4.5
Dwingeloo 1	8.4
Einstein Cross	2.7
ESO 202-23	10.4
ESO 510-13	1.25
G1	8.6
Hoag's Object	1.28
Holmberg II	8.5
IC 1296	10.3
IC 1308	4.3
IC 2233	8.12
IC 2574	5.3
IC 4107	3.2
IC 5146	1.6
II Zw 40	1.23
II Zw 99	9.1
M 31	1.4
M 33	7.7
M 51	7.2
M 67	5.2
M 77	1.27
M 81	5.5
M 82	1.21
M 83	3.3
M 101	4.4
M 104	7.1
M 108	6.8
Malin 1	1.32
MCG chain	4.5
NGC 55	1.12
NGC 100	1.26
NGC 128	1.19
NGC 185	8.1
NGC 247	7.6
NGC 253	3.1

Object	Fig.
NGC 300	7.5
NGC 404	10.1
NGC 660	6.7
NGC 891	6.6
NGC 1049	8.2
NGC 1232	1.13
NGC 1365	1.20
NGC 1531/32	7.5
NGC 1618/22/25	10.2
NGC 2573	7.14
NGC 3172	7.13
NGC 3314	2.5
NGC 3628	7.4
NGC 3877	8.7
NGC 4038/39	1.35
NGC 4319/Mrk 205	2.6
NGC 4449	7.9
NGC 4486B	1.24
NGC 4517	3.6
NGC 4565	1.14
NGC 4622	1.34
NGC 4631/27	3.8
NGC 4650A	3.5
NGC 4676	8.15
NGC 5090/91	1.1
NGC 5128	7.8
NGC 5266	1.18
NGC 5421	3.7
NGC 5907	8.11
NGC 6166	1.22
NGC 6240	1.31
NGC 6726/27/28	1.7
NGC 6745	8.13
NGC 6946	1.16
NGC 7331	7.10
OM-076	2.8
PG 1634+706	5.4
PGC 61965	8.8
PKS 2349–014	1.36
Scultpor Dwarf	1.30
Seyfert's Sextett	9.3
Shkh 1	2.9
UGC 3697/3714	7.11
UGC 10214	8.14
UGC 12914/15	9.2
V362 Vul	8.9
WLM	8.3
3C 66A	6.5
8 Zw 288	9.4

Sources of Figures

Designation	Meaning	Figures
AAO	Anglo Australian Observatory	2.3
Bresseler	Peter Bresseler	1.18, 1.22
Cardiff	University of Cardiff (Jonathan Davies)	1.32
Celnik	Werner E. Celnik	6.1
Chandra	Chandra X-ray Center, CfA, Harvard	2.13
DSS	Digital Sky Survey, Space Telescope Science Institute	1.18, 1.19, 1.23, 1.24, 1.26, 1.30, 1.31, 2.8, 2.9, 3.2, 3.6, 3.7, 3.9, 5.2, 5.3, 5.4, 6.5, 7.9, 7.11, 7.13, 7.14, 8.2, 8.5, 8.8, 8.9, 8.12, 9.1, 9.2, 9.4, 9.5, 9.6, 9.7, 10.1, 10.2, 10.3, 10.4
ESO	European Southern Observatory	1.1, 1.13, 1.20, 3.3, 7.1, 7.5, 10.8
Flach-Wilken	Bernd Flach-Wilken (www.spiegelteam.de)	1.14, 6.6, 7.4
Güths	Torsten Güths	4.2
HST	Hubble Space Telescope, Space Telescope Science Institute	1.25, 1.28, 1.33, 1.34, 1.35, 1.36, 1.37, 2.5, 2.6, 2.7, 2.14, 2.16, 8.6, 8.13, 8.14, 8.15, 9.3
ING	Isaac Newton Group	1.16, 1.27, 5.5, 7.10, 8.1, 8.3, 8.4
Keck	W. M. Keck Observatory	2.12
Koch	Bernd Koch	2.10, 8.7
Maddox	University of Nottingham (Steve Maddox)	3.4
Poyner	Gary Poyner	6.2
Scholz	Cord Scholz	1.2, 1.4, 1.6, 3.8, 4.4, 7.9, 8.11
SDSS	Sloan Digital Sky Survey	1.21, 1.22, 2.11, 4.5, 6.7, 7.2
Sparenberg	Rainer Sparenberg	1.5, 1.12, 3.1
Virgo C.	Virgo Consortium, MPI for Astrophysics	2.4
Wendel	Volker Wendel (www.spiegelteam.de)	1.7, 7.3, 7.6

All other figures (images, sketches) are from the author (adapted from various sources) and maybe changed by Springer.

Printed in Singapore